PRENTICE HALL

Language Teaching Methodology Series

Classroom Techniques and Resources

General Editor: Christopher N. Candlin

Video in Action

Other titles in this series include

CANDLIN, Christopher and MURPHY, Dermot
Language learning tasks

ELLIS, Rod
Classroom second language development

ELLIS, Rod
Second language acquisition in context

FRANK, Christine and RINVOLUCRI, Mario
Grammar in action

GOLEBIOWSKA, Alexandra
Getting students to talk

KENNEDY, Chris
Language planning and English language teaching

KRASHEN, Stephen
Second language acquisition and second language learning

KRASHEN, Stephen
Principles and practice in second language acquisition

KRASHEN, Stephen
Language acquisition and language education

KRASHEN, Stephen and TERRELL, Tracy
The natural approach

MARTON, Waldemar
Methods in English language teaching: frameworks and options

McKAY, Sandra
Teaching grammar

NEWMARK, Peter
Approaches to translation

NUNAN, David
Understanding language classrooms

PECK, Antony
Language teachers at work

ROBINSON, Gail
Crosscultural understanding

STEVICK, Earl
Success with foreign languages

SWALES, John
Episodes in ESP

TAYLOR, Linda
Teaching and learning vocabulary

WALLACE, Catherine
Learning to read in a multicultural society

WENDEN, Anita and RUBIN, Joan
Learner strategies in language learning

YALDEN, Janice
The communicative syllabus

Video in Action

Recipes for Using Video in Language Teaching

SUSAN STEMPLESKI
(Hunter College, New York)

BARRY TOMALIN
(BBC English, London)

ENGLISH LANGUAGE TEACHING

Prentice Hall

New York London Toronto Sydney Tokyo Singapore

First published 1990 by
Prentice Hall International (UK) Ltd
66 Wood Lane End, Hemel Hempstead
Hertfordshire HP2 4RG
A division of
Simon & Schuster International Group

Typeset in 10½/13 pt Times
by Keyset Composition, Colchester

Printed and bound in Great Britain at the
University Press, Cambridge

Library of Congress Cataloging-in-Publication Data

Stempleski, Susan.
Video in action: recipes for using video in language teaching/
Susan Stempleski and Barry Tomalin.
p. cm. — (Language teaching methodology series)
Includes bibliographical references.
ISBN (invalid) 0–13–946519–8 : $12.00
1. Languages, Modern—Study and teaching. 2. Video tapes in education. I. Tomalin, Barry, 1942– . II. Title. III. Series.
PB36.S7 1990
418′.0078—dc20 89-16269
CIP

British Library Cataloguing in Publication Data

Stempleski, Susan 1941–
Video in action : recipes for using video in language teaching. (Language teaching methodology series)
1. Languages. Teaching. Applications of educational technology
I. Title II. Tomalin, Barry 1942–
407′.8

ISBN 0–13–945619–6

1 2 3 4 5 94 93 92 91 90

Contents

General Editor's Preface

Within the Language Teaching Methodology Series we have created a special set of books with the . . . *In Action* title. These books are designed to offer teachers material that can be directly used in class. They are resources for action, hence the title. The books we will be increasingly publishing in this sub-series will be characterised by a wealth of usable classroom activities, all of which will have been selected and presented on the basis of having being tried and tested by practising teachers. The books are designed to be open, with easily accessible tasks for teachers and learners. More than this, they are constructed to reflect recent results from research and to have built into them the potential for teacher self-development and for in-service teacher education and training. They are thus both for the learner and for the teacher.

Susan Stempleski and Barry Tomalin have created a collection of techniques and recipes for the imaginative use of video in the classroom, drawing on ideas from their own wide experience in many countries, and on other specialists' experiments in their classrooms. Intended for all teachers of languages and not just for EFL/ESL, the book offers ways to motivate language learners with varying levels of competence and to create possibilities for enhancing classroom communication.

Non-verbal behaviours are among the hardest to make learners aware of, yet we know their significance for communication, expecially cross-culturally. There are many ideas here that offer teachers principled ways of exploiting that undertaught mode of communication, especially in combination with more conventional skills practice.

The increasing emphasis on content-based curricula for secondary and primary language learners as well as for adults naturally encourages the use of video material. Adapting and exploiting that material for the classroom is not always easy to manage, however, especially for hard-pressed classroom teachers. The authors have taken this into account in designing the book, and, of course, the sub-series as a whole has these logistic and pedagogic problems very much in mind. Accordingly, of particular value will be the way the book is constructed to help teachers plan the place of video in their curricula, evaluating its advantages over other media (and in combination with them) and indicating how it can be both the focus for and an important ancillary to the classroom lesson.

Finally, and in keeping with the . . . *In Action* sub-series, the book itself is constructed to be maximally accessible and easy to use. Different ways of referencing the material have been used to make planning easier for the teacher, incidentally offering valuable ideas about how to catalogue and store teaching and learning resources.

Professor Christopher N. Candlin
General Editor
Macquarie University, Sydney

Acknowledgements

For a book of this kind, it is impossible to trace the exact origin of each idea or technique. The authors would therefore like to acknowledge, with grateful thanks, the following people, whose work has served as an inspiration for the compilation of the recipes in this book:

Margaret Allan
Paul Arcario
Patricio Bracamonte
Leslie Dunkling
John Fanselow
Marion Geddes
Joan Giummo
David Haines
Meryl Harris
Brian Hill
Susan Kellerman
Mike Lavery
Jack Lonergan
Bob Marsden
Pamela McPartland
Keith Morrow
Janet Moser
Hamish Norbrook
Mario Rinvolucri
Consuelo Rivera
Marita Schocker
Gill Sturtridge
Anthea Tillyer
James Willimetz
Jane Willis

Particular thanks to Joe Hambrook for his suggestions and painstaking editorial work in ironing out some of the inconsistencies. Any that still remain are the authors'.

INTRODUCTION

Introduction

Video in Action is a *double* first for language teachers.

It's the first full-scale collection of recipes for using video in class, pooled from the ideas and experiments of experts and enthusiasts world-wide. Having chosen a video sequence, you can browse through the recipes until you find one that takes your fancy. Or you can refer to the indexes for specific activities and ways of treating different kinds of video to suit different classroom situations. This book will also help you to deal with equipment and run a video library, and it answers a number of questions about video and its use.

Video in Action is also the first publication to combine British and American experience in using video for language teaching. Among the many benefits of this collaboration is the extended range of activities for exploiting authentic TV broadcasts, advertising commercials, music videos and feature films, as well as basic guidance in the use of video. And although the emphasis is on the teaching of English as a second or foreign language, *Video in Action* is equally suitable for other target languages.

How video can help

From their wide experience in training teachers in primary, secondary and adult education world-wide, the authors are convinced that the introduction of a moving picture component as a language teaching aid is a crucial addition to the teacher's resources.

First of all, through *motivation*. Children and adults feel their interest quicken when language is experienced in a lively way through television and video. This combination of moving pictures and sound can present language more comprehensively than any other teaching medium. And more realistically too. Using a video sequence in class is the next best thing to experiencing the sequence in real-life. In addition, video can take your students into the lives and experiences of others. Through the recipes in *Video in Action* you can see how best to exploit this motivation and guide it into successful language learning.

Second, through *communication*. Teachers have observed how a video sequence used in class makes students more ready to communicate in the target language. This book contains many communication activities which encourage students to find out things from each other on the basis of the video itself. Some of these activities depend on information gaps, created by manipulating the technology or the viewer so that an individual viewer can get the full message only by communicating with another viewer. Other activities depend on the richness of content in TV programmes, feature films, commercials, etc., and give students the opportunity to conduct opinion polls and consumer surveys in class. By such means *Video in Action* encourages a more interactive classroom.

Third, through *non-verbal aspects* of communication. The American psychologist, Robert Merabian, has estimated that as much as 80 per cent of our communication is non-verbal. Our gestures, expression, posture, dress and surroundings are as eloquent as what we actually say. Video allows us to see this in action and to *freeze* any moment to study the non-verbal communication in detail. Recipes like 'Body language' and 'In the mood' offer suggestions for exploiting this unique contribution of video to language learning.

Finally, through *cross-cultural comparison.* Observing differences in cultural behaviour is not only suitable training for operating successfully in an alien community. It is also a rich resource for communication in the language classroom, which recipes such as 'What if . . .' and 'Culture comparison' show you how to exploit.

When to use video

It is only recently that video has moved from being something that is switched on and left to present language without the teacher's intervention, to becoming a resource for classroom activities in which many different kinds of video material are exploited in a variety of ways. *Video in Action* carries this movement even further by offering detailed suggestions for choosing, preparing and exploiting more than seventy different video sequences in class. Most of the sequences are short (some of them lasting less than half a minute) and generally easy to obtain.

Video can be used at every level, both as supplementary material for language reinforcement and skills practice (ideally once a week but at least once every two weeks) and as the main component of an intensive course or course module provided that suitable material is available.

Apart from intensive video-based courses, it is probably best for institutions to give beginners and elementary-level students priority in the use of video, since they are likely to benefit most from its power to motivate and to provide direct access to comprehension by putting language in context.

In the end, the use of video depends on the resources and time which an institution and its teachers can devote to it. But even at the most modest level, the recipes here will help them exploit video to its utmost effect.

How to use this book

The Indexes

Your key point of reference is the Indexes. The first part lists the recipes alphabetically with a note on their three main characteristics or categories (level, purpose and sequence type). The second part lists them according to each of these categories in turn. This information will enable you to match your video sequences with suitable classroom activities.

Here is how you use them:

1. *Index of recipes:* this lists all the recipes in the alphabetical order in which they appear in the book, so you can see at a glance which may suit your particular requirements.
2. *Index to level, purpose and sequence type:* this index will help users gain easy access to suitable recipes by three possible routes:

   - Level – your students are, for example, elementary level and you want to see what recipes will be suitable for them.
   - Purpose – you want to use video in, for example, vocabulary development or grammar review and you want to see what recipes will be suitable.
   - Sequence type – you have, for example, a pop music video or a feature film and you want to see what recipes will be suitable.

The main categories

The main categories are indicated at the beginning of each recipe.

Level

Most recipes are suitable for students at the level indicated and for those above that level. The levels are:

1. *Beginners:* students with no prior knowledge of the target language.
2. *Elementary:* students who have had some contact with the target language (e.g. as false beginners or by having studied it for several weeks in a regular course).
3. *Intermediate:* students who have a basic knowledge of tenses and structures and an ability to communicate with other students at a basic level or beyond.

4. *Advanced:* students with a proficiency or post-intermediate command of the target language and an ability to speak it and write it with some fluency.
5. *Children:* students from approximately 8 to 12 years old.

Purpose

Most video sequences can serve more than one purpose, depending on the teacher's objectives for the lesson. The recipes usually list more than one purpose to suit the recommended activities. The different types of purpose are:

1. *Active viewing.*
   - It is essential to engage students *actively* whenever they watch video material, and also to introduce them to the content of each sequence before studying the language by giving them specific viewing tasks. This active involvement should form part of the early stage of any lesson with video, but it has been specifically mentioned in recipes where it is particularly important.
2. *Vocabulary.*
   - Vocabulary development – activities focusing on new lexical sets to be learnt through the sequence.
   - Vocabulary review – activities reinforcing language already presented or learnt.
3. *Grammar.*
   - Grammar presentation – activities for presenting particular grammatical structures.
   - Grammar review – activities reinforcing grammatical structures.
4. *Pronunciation.*
   - Activities focusing on sounds, stress and intonation.
5. *Listening/speaking skills.*
   - Viewing comprehension – activities focusing on the visual element in the sequence.
   - Listening – activities focusing on what is said in the sequence.
   - Oral composition – activities focusing on telling the story of the sequence orally.
   - Speaking – activities focusing on structured presentations related to the subject of the sequence.
   - Discussion – activities focusing on discussions about the sequence.
6. *Reading/writing skills.*
   - Reading – comprehension activities based on reading materials related to the sequence.

- Note-taking – activities teaching or practising note-taking skills.
- Written composition – activities focusing on creative writing and/or summarising the sequence.

7. *Cross-cultural concerns.*
   - Cultural awareness – activities focusing on studying the cultural influences at work in the sequence.
   - Cross-cultural comparison – activities focusing on the comparison between the culture of the mother tongue and the target culture.

8. *Testing.*
   - Activities which can be used as a testing format based on the sequence.

Sequence type
The most obvious division between types of video sequences is that which separates educational videos made specifically for language learning and *authentic* video material (i.e. video/TV/film material made originally for native speakers). The recipes in *Video in Action* treat both types equally without distinguishing between them.

The general categories below include subcategories, as indicated, and several overlap each other.

1. *Drama:* this broad category embraces most video, TV and film material with scripted dialogue and/or dramatic elements, such as TV soap operas, drama series, plays, situation comedies, etc. It also includes feature films as a subcategory.

2. *Documentaries:* any material which is non-fictional and/or unscripted (e.g. a TV current affairs investigation of an issue as opposed to a scripted drama about the same issue). Subcategories include TV news programmes, interviews and sports programmes, and also the kind of TV talk shows indicated in some of the recipes.

3. *TV commercials:* all kinds of TV or cinema advertising material or short promotional videos (except music videos).

4. *Music videos:* generally used to refer to pop music videos of all kinds, but could include songs and musical numbers from TV light entertainment shows or cinema musicals.

Other categories

In addition to the main categories each recipe includes the following.

Sequence length
This suggests a suitable length of video sequence for the activities in the recipe concerned, often within quite wide time limits. It is difficult to recommend an exact length without actually identifying a particular video sequence, but the suggestions are based on the authors' own successful experience.

In general, it is better to exploit a short segment of video thoroughly and systematically rather than to play a long sequence which is likely to result in less active viewing on the part of your students.

A word of warning: when you use short isolated sequences, you must expect your students to be interested in other parts of the video concerned, and you should be prepared to respond to this interest (e.g. to play and eventually exploit the whole of an episode or the serial of which it forms part, not just the sequence you have chosen for one of the recipes in this book).

Activity time
This suggests a suitable length of time for each recipe but, as with sequence length, it is only a rough guide which you may need to modify depending on the language level of your students and the make-up and size of your class.

You may also want to shorten or extend certain activities, or to replay a video sequence several times, or to use two or more recipes in a single lesson period. Indeed, short activity times are often suggested so that you can combine several recipes in this way, as suggested in 'Stop/start' (see page 108).

Preparation

In each recipe, *Video in Action* offers some basic guidance for choosing suitable video sequences and preparing other materials. Although it does not give many details, this advice is intended to draw your attention to the importance of preparing your video-based lessons thoroughly. Here are some general hints, especially for those who are inexperienced in using video.

General considerations
Preparing a lesson based on a video sequence can be difficult and time-consuming. The pay-off is that the same lesson can be used again and again by many teachers. So three hours spent in preparing a single lesson can provide very many hours of teaching.

Published language-teaching video material is accompanied by textbooks which usually include guidance for teachers. Indeed, many of the better video-based courses integrate the video component into a complete multi-media package. However, even if you are using a published course, you may want to produce your own lesson plans to fit your timetable and the specific needs of your students. And if you are planning to use authentic video material or to use language teaching video as supplementary material, you will have a lot of preparation to do.

Video as text

Treat each sequence as a text, just like a language presentation passage in a book or a dialogue on audiocassette. Plan your lesson using *both* the video script, if available, *and* the video itself. The script will tell you what language is used; the video will provide essential evidence on behaviour, character and context, which are not usually in the script.

Selecting a sequence

Assuming that you have a fairly wide range of video material available, the following factors should influence your choice of a particular sequence for use in class.

Interest: Students don't think of video as *teaching* material. They think of it as television. Therefore, if they consider that a sequence is dull, they won't want to watch it or learn through it. Any sequence chosen for use in class must be intrinsically interesting or attractive and must comprise a complete unit of meaning regardless of its context.

Length: With a few exceptions, your video sequences must be suitable for exploitation in a single one-hour class period. They are therefore likely to be no more than five minutes in length (and may be as short as thirty seconds) for most classroom activities.

Flexibility: Most sequences should be suitable for *several* of the activity categories described in this book.

Language level: This is not necessarily a crucial criterion. It is more important to grade the task or activity so that the class can deal with it, rather than to grade the video material itself. In many instances, the picture offers clues to meaning over and above the language on the soundtrack. Thus, much apparently difficult authentic video material can be used with students at a relatively low level of ability.

Language items to be taught: If the aim of the lesson is to teach language from the screen, then the presence of relevant functions and structures will be a prime consideration in selecting a sequence. If, however, the aim is to stimulate discussion or other skills work among the students, then specific language items are not so important.

Lexis: Though video can be used very effectively to introduce and review vocabulary (see Indexes), this is seldom the most important criterion for selecting a particular sequence.

Selecting recipes and activities
Go through the index of recipes and find one or more suitable recipes for exploiting your chosen sequence with the particular class concerned. In addition, you will need to answer the following questions:

- What language (if any) needs *pre-teaching*? Ideally, there should be none at all. The students should be able to ask for help with new language, which should be part of the communicative activity.
- What *viewing task* should you choose to focus the students' attention on the content of the sequence? Each recipe recommends one or more tasks, but you may wish to vary these.
- Do you need a *sequencing task* to help the students fix the order of events in mind?
- What *language* from the sequence do you want to teach?
- What *follow-up* activities do you want to use?

Preparing worksheets, documents, presentations, etc.
You will often need to prepare worksheets and blackboard or overhead projector presentations for a variety of purposes including viewing and sequencing tasks, language review and consolidation, roleplays and other extension activities.

In addition, you will sometimes need to prepare transcripts of dialogues, commentaries, etc., from the video sequences you are going to use. This work can be extremely time-consuming.

Planning the use of time
Unless your institution is particularly well-endowed with video equipment, you will have to fit your video-based lessons into times when the equipment is available, and this may mean adjusting the planning and scheduling of your lessons as a whole. At the same time, you should plan so that, as far as possible, only work essential to video use is done during the video lesson period, while preparatory and follow-up activities are allocated to times when the video equipment is not available.

The ideal length for the video lesson itself is between 45 minutes and one hour, but time may also have to be allowed for setting up the video and moving equipment. And if your institution has a special video room, your class will need extra time to get there and back.

Sharing the workload, updating, etc.
You may have to prepare your video lessons entirely on your own, but if possible you should share the work with other colleagues who use video, maybe even those working in another institution.

Frequent discussion with colleagues who use video, comparing notes and experience, amending and updating lesson plans and materials together, can enormously increase the effectiveness of your video-based work. For this purpose you should keep notes on the individual recipes you use (e.g. on the sequences which work best with a particular recipe). Try to use a word processor for such documents, and for transcripts of videos, so that they can be updated easily.

In class

The recipes in this book give detailed step-by-step suggestions for the conduct of lessons involving video. However, you should treat these as suggestions, not instructions, and vary them to suit your students' needs and abilities, and, of course, to suit the video sequences you have selected.

Above all, lessons with video should be enjoyable and should provide even mediocre students with a genuine sense of achievement: when, for instance, they have grasped the meaning of quite difficult dialogue because they could follow the action and behaviour which accompanied it; or when they have succeeded in predicting the outcome of a story on the basis of a few visual clues; or when they can use new language effectively, having seen it used in a clearly meaningful context on video.

Basic techniques

Several of the recipes use basic techniques for exploiting video. As a quick reference guide and for teachers who have little or no experience in using video, a selection of these basic techniques has been extracted and can be found at the beginning of the Recipe section of the book.

The Appendices

Some of the main concerns for schools and institutions where video is used are about logistics rather than methodology. Appendix 1 deals with basic features of video equipment, the running of a video library or resource centre, the maintenance of videocassettes and questions concerning copyright.

There are two simple checklists: one for all teachers using video in class, and one for those about to use video for the first time.

Appendix 2 gives a current analysis of the different TV standards in use around the world.

THE RECIPES

Some basic techniques for using video

The recipes in this section of the book include all of the basic techniques, applied to a variety of different situations and activities, and many more besides. Below is a selection of these basic techniques for less-experienced teachers to familiarise themselves quickly with some of the things they can do with video before trying the recipes.

1. *Sound off/vision on (silent viewing):* this techniques can be used either to stimulate language activity about what is seen on the screen (rather than what is being said) or to focus on what is being said, by a variety of guessing/prediction tasks. The most common ways of using this technique are:

    - Choose scenes with short exchanges of dialogue, where the action, emotions, setting, situation, etc. (even lip movements!), give clues to what is being said. Students guess/predict the words and then compare afterwards while viewing with sound on.
    - Use longer exchanges for students to guess the gist or situation rather than exact words.
    - Play whole sequences for students to try and write suitable screenplays, which can then be compared with the actual soundtrack.
    - Use the pause/freeze-frame control at the initial point of each exchange, for students to predict language on a line-by-line basis. These are then compared with the actual speeches.
    - Students give or write a descriptive commentary of what they see.

2. *Sound on/vision off:* students guess the setting, action, characters, etc., from the soundtrack. This can be done in a variety of ways as in item 1 above, i.e. on a line-by-line or scene-by-scene basis.

3. *Pause/freeze-frame control:* (to utilise with sound on/off and vision on/off controls as above.)

    - With sound on, pausing at the initial point of each exchange, teacher asks students to predict the words. Immediate comparison with the actual words can then be made by playing each exchange.

- With sound on, pausing at strategic points in the plot/action, teacher asks students questions about the situation (what has happened/what is going to happen).
- Pause at suitable moments of characters' facial expressions for students to suggest thoughts, feelings, etc.

4. *Sound and vision on (listening and viewing comprehension):*

   - Students are given a list of items *before* viewing a sequence and have to look for them as they view.
   - Students are given a list of items *after* viewing a sequence and have to decide which were in the sequence and which weren't.
   - Students view and listen to the sequence for (e.g.) something beginning with (X), something (blue), something which rhymes with (Y), etc.
   - Students are given comprehension questions before viewing a sequence and answer the questions after viewing.
   - Students are told what a sequence will be about and have to list the things they expect to hear, see, etc. After viewing, they can then compare and discuss.
   - Students are given paraphrases of dialogues before viewing and then have to match/spot the direct speech as they view.
   - Students are given a cloze passage of the dialogue or of a description of the scene and have to complete gaps while/after viewing.

5. *Jumbling sequences:* students view each section of a sequence, presented to them *out of sequence.* They have to determine what has happened/what will happen in each case and then fit the sections into a correct or plausible sequence.

6. *Split viewing:* some students see a sequence but do not hear it; others hear but do not see. A variety of activities can then follow based on usual information-gap procedures.

A change of seasons

Level Intermediate and above

Purpose Discussion, Speaking

Sequence type Drama

Sequence length 1–3 minutes

Activity time 10–15 minutes

Preparation

Select a sequence in which the season or time of year concerned has an important effect on the story, setting, characters, etc.

In class

1. Tell the students that you are going to play a sequence which takes place in a particular season or time of year. Their task is to imagine how the sequence would be different if it took place during another season or time of year.
2. Write the following questions on the board:

   If this scene took place in [season], what differences would there be in:
   - *What the characters say?*
   - *What the characters are wearing?*
   - *What the characters do?*
   - *The story or situation as a whole?*
3. Play the sequence.
4. The students work in groups of three or four discussing the questions.
5. Groups report to the whole class.
6. Play the sequence again, as a final check on the effect that the season/time concerned has on the characters, story, etc. Continue the discussion if required.

Adjective match

Level Intermediate and above

Purpose Speaking, Vocabulary development

Sequence type Drama

Sequence length 1–3 minutes

Activity time 15–20 minutes

Preparation

Select a sequence featuring three to five characters with different personality types. Prepare a list of 15–20 adjectives describing character traits people can have (e.g. cheerful, gloomy, helpful, selfish, friendly, unfriendly, trusting suspicious, naïve, sophisticated, etc.).

In class

1. Distribute the list and pre-teach any new words.
2. Tell the students that you are going to play a sequence in which each of the characters could be described by at least one of the adjectives on the list. Their task is to match up each character with one or more of the adjectives.
3. Play the sequence.
4. The students work in groups of three or four, discussing what the characters do and say during the sequence and matching up each character with at least one of the adjectives.
5. After the group discussions, individual volunteers choose one of the adjectives and tell the class which character they think it describes. They then give an example of behaviour demonstrating the meaning of the adjective.

Analysing commercials

Level Intermediate and above

Purpose Discussion, Listening, Note-taking

Sequence type TV commercial

Sequence length 30–60 seconds

Activity time 30 minutes

Preparation

Select one or more commercials which provide enough relevant information and discussion points for this activity. Duplicate the chart on page 21, with a copy for each student.

In class

1. Distribute the chart. Go over it with the students to make sure they understand the kind of information required.
2. Tell the students that you are going to play a TV commercial. Their task is to complete the chart with information from the commercial.
3. Play the commercial, several times if necessary.
4. The students work individually to complete the chart.
5. As they finish, ask students to compare their answers with those of another student.
6. Play the commercial again. The students confirm or modify their answers.

Variation

If time is short, divide the class into groups. Each group looks for information to complete one section of the chart. At the end of the activity the groups compare notes.

Discussion

The students analyse two or three commercials in this way and then compare them.

TV commercial information sheet

Product information:

Brand/company name ____________________

Product/service ____________________

About the product ____________________

About the seller ____________________

Consumer information:

About the buyer ____________________

About the user (may be the same as the buyer) ____________________

Other information (concerning changes the product will make in the life of the buyer/user)

____

Language used:

'Plus' words ____________________

'Minus' words ____________________

Action words ____________________

Made-up words ____________________

Asking real questions

Level	Intermediate and above
Purpose	Grammar review (question forms), Listening, Speaking
Sequence type	Documentary, TV news programme
Sequence length	3–5 minutes
Activity time	15–20 minutes

Preparation

Select a documentary sequence or news item which contains information about a subject of interest. News items introduced by an on-screen presenter are excellent (CNN or CBS News, for example).

In class

1. Ask for volunteers to say what they know about the subject treated in the sequence. Write their contributions on the board.
2. Tell the students that they probably have a lot of questions about the subject. Volunteers ask one or two questions.
3. The students work in groups of three or four and write down at least three other questions about the subject.
4. Tell the students that you are going to play a sequence in which some of their questions may be answered. The questions may be answered directly, by inference or not at all. Their task is to determine what answers, if any, are given to their questions.
5. Play the sequence.
6. Volunteers read their group's questions to the class and report on answers provided by the video.
7. The remaining questions for which no answers have been provided can form a topic for project work or library research.

Variation

Play the opening of the sequence. Ensure that the students understand the topic. Elicit questions and write them on the board before playing the rest of the sequence. This variation is especially successful when the sequence is introduced by a presenter speaking to camera.

Assemble the script

Level	Intermediate and above
Purpose	Listening, Speaking, Viewing comprehension, Written composition
Sequence type	Drama
Sequence length	30 seconds to 2 minutes
Activity time	30 minutes plus

Preparation

Select a sequence in which the dialogue provides several clues to the action, and the picture frequently suggests what is being said. You will need two rooms and an audiocassette recorder. Before class, record the soundtrack of the sequence onto an audiocassette.

In class

1. Divide the students into two teams and possibly into subgroups.

2. Tell the students that you are going to play a short sequence. Explain that one team will have the soundtrack only: they must imagine the pictures. The other team will have the video without sound: they must write the dialogue script. If necessary, give a very brief hint about the subject-matter of the sequence, the names of characters, etc.

3. Team 1 takes the audiocassette recorder to the other room. They play the soundtrack and write down what they think the *situation* is, who the *characters* are, and *what happens* during the sequence.

4. Stay with Team 2. Play the complete sequence with the sound turned down. Then play it shot by shot without sound, pausing to allow the team to write the dialogue.

5. Bring Team 1 back into the classroom. Divide the students into pairs with one member from Team 1 working with one member from Team 2.

6. Each pair takes a piece of paper with a line down the middle. They must now write the script (short description on the left of the line, dialogue on the right). For example:

Man walks into the room.	*MAN:*
Woman is sitting there.	*Hello, Darling, I'm home.*

   Go round checking the script:

7. Several pairs read their scripts to the class.

8. Play the sequence again as a check.

Biographies

Level Intermediate and above

Purpose Written composition

Sequence type Drama

Sequence length 5–10 minutes

Activity time 20–30 minutes

Preparation

Select a sequence featuring several characters in ways that will encourage speculation about the characters themselves, their origins, background, etc.

In class

1. Write the following questions on the board:

   What is the character's full name?
   Where was he or she born?
   When was he or she born?
   What were his or her parents (or family) like?

2. Tell the students the general nature of the sequence you are going to play. Their task is to choose one of the characters, and then write a brief, imaginary biography about him or her, using the questions above as a starting-point. They may include information from the sequence itself.
3. Play the sequence.
4. The students individually write the biographies within a set time limit.
5. When the time is up, the students work in small groups, taking turns to read their biographies to the group.
6. Play the sequence again, and look for further information which could confirm or modify the students' biographies.

Variation

If your students are fairly advanced, you may wish to have them view a relatively long scene (or an entire film) and then write biographies of the character for homework.

Body language

Level Intermediate and above

Purpose Note-taking, Testing, Viewing comprehension

Sequence type Documentary, Drama

Sequence length 30 seconds to 3 minutes

Activity time 20 minutes

Preparation

Select a sequence with social interaction in which the speakers' actions don't necessarily back up what they are saying (e.g. verbally they co-operate, but their actions show anxiety, lack of interest, etc.).

In class

1. Warm up by getting individual students to show socially acceptable gestures, facial expressions or postures. Other students must guess what they mean.
2. Write the following words on the board and explain them:

   HANDS *(gestures)*
   EYES *(eye contact)*
   MOVEMENTS *(posture/movement)*
   FACE *(happy/unhappy/serious/cheerful/etc.)*

3. Tell the students that you are going to play a sequence with several characters. Their task is to choose one character and list the things he or she does with hands, eyes, etc. Identify the characters so that the students can choose one before viewing the sequence.
4. Play the sequence. The students make notes.
5. Elicit information about characters' body language. Write it on the board.
6. Play the sequence again and pause at body language points. The students discuss the real attitude of the characters.

Variations

■ Matching phrases to body language

In this variation, choose phrases to express each gesture, movement, etc (e.g. (a) I don't care, (b) I'm frightened). Make sure the students understand each phrase. Play the sequence without sound. Pause at each body language point. The students match a phrase to the body language. At the end, play the sequence with sound and vision. Discuss the students' interpretations of the body language.

■ Cross-cultural differences

Discuss differences between body language in the video and body language with similar meaning(s) in the students' own culture(s). Identify and discuss other significant body language variations between cultures.

Catch the credits

Level Elementary and above

Purpose Reading, Viewing comprehension, Written composition

Sequence type Feature film

Sequence length 1–3 minutes

Activity time 5–10 minutes

Preparation

Select a sequence from a feature film which shows the opening title and film credits. Prepare a list of between five and ten questions about things written on the screen during the sequence (e.g. What is the title of the film? Who wrote it? Who directed it? Who produced it?). Make enough copies of the list to give one to each student (unless you are going to write it on the board).

In class

1. Distribute the list, or write it on the board.
2. Explain to the students that you are going to play the opening credits to a film. Their task is to write down the answers to the questions.
3. Play the sequence, more than once if necessary. Students note down their answers.
4. Students work in pairs, comparing their answers and writing sentences containing the information in the credits, e.g. '*Witness* was directed by Peter Weir'.
5. Play the sequence again, pairing after each answer is shown on the screen. Volunteers take turns reading their sentences to the class. Write the sentences on the board. Elicit necessary corrections in grammar and content from the class.

Homework

Students write a paragraph summarising the information contained in the opening credits.

> NOTE: *Title sequences and opening credits are frequently associated with visual material and sound meant to reveal or suggest aspects of the film. If this is the case with the sequence selected, you might wish to follow up this activity with a Prediction activity (see Indexes).*

Celebrity interviews

Level Intermediate and above

Purpose Grammar review, Note-taking, Speaking, Viewing comprehension

Sequence type Drama, Documentary, Interview, Music video, Talk show, TV news programme

Sequence length 2–3 minutes

Activity time 20–30 minutes

Preparation

Select a sequence which features a celebrity or someone worth interviewing (e.g. because he or she has done something remarkable or witnessed a unique event).

In class

1. Tell the students the general nature of the sequence you are going to play. Their task is to focus on a particular person or celebrity, and later make up a list of questions to ask the interviewee. Either identify the person concerned or tell the students to choose someone from the sequence.
2. Play the sequence, pausing from time to time to allow students to take notes.
3. The students work in groups, preparing an interview of the chosen person. Ask them to make up a list of actual questions they would use to interview him/her.
4. Ask for a volunteer to be the chosen celebrity or person. Groups take turns asking their questions. If the language level of the class is high enough, the volunteers can be encouraged to answer.

Variation

The event can be made more dramatic by setting the interview up as a press conference with a chosen celebrity or person and several interviewers.

Colour game

Level Elementary, Children

Purpose Vocabulary development, Vocabulary review

Sequence type Any

Sequence length 30 seconds to 1 minute

Preparation

Select a suitable sequence. It must include colours that can be identified and possibly talked about. Prepare identification tasks as suggested below.

In class

1. Divide the class into four groups.
2. Tell the students/children that you are going to play a sequence which features several colours. Give each group an identification task such as, 'Tell me in English the first green thing you see.' 'Who is wearing a red hat?' 'Look for something red and something blue.' Some groups will have the same tasks as others.
3. Play the sequence.
4. Each group decides on the answer to their task. If necessary they consult you or a dictionary.
5. Elicit responses from the groups.
6. Play the sequence again and check.

Complete the story

Level Elementary and above

Purpose Grammar review, Reading

Sequence type Drama

Sequence length 3–5 minutes

Activity time 10–15 minutes

Preparation

After selecting a suitable story sequence, prepare a synopsis omitting key words for the students to fill in (e.g. every seventh word, or another variant). Make a copy for each student.

In class

1. Distribute copies of the synopsis.
2. Tell the students that you are going to play the sequence summarised in the synopsis. Their task is to supply the missing words.
3. Play the sequence.
4. The students work in pairs or small groups, writing in the missing words.
5. Volunteers take turns reading parts of the synopsis and their suggestions for the missing words. List their suggestions on the board. Wherever possible, encourage the rest of the class to offer alternatives to these suggestions.
6. Play the sequence again to check the suggestions and alternatives against the video.

Variations

- Instead of having the students write after viewing the sequence, proceed directly to the reading phase and ask students to supply the missing words orally.
- Before viewing, the students predict the missing words and then check them when the sequence is played.

Consumer survey

Level	Intermediate and above
Purpose	Discussion, Viewing comprehension, Written composition
Sequence type	TV commercial
Sequence length	30–60 seconds
Activity time	30 minutes

Preparation

Choose a commercial for a product or service which is likely to be relevant to the consumer interests of the class. Duplicate the questionnaire on page 37, with enough copies for all the students.

In class

1. Distribute the questionnaire. Tell the students to interview each other and answer the questions. Every student should complete the questionnaire.
2. The students report back to you and compile a consumer profile of the class (e.g. average number of hours of TV watched per day, per week, etc.).
3. Tell the students that you are going to play a TV commercial. Write the following questions on the board and make sure that the students understand them: 'What is the message?' 'At what target audience is the message aimed?' 'Are you part of the target audience?' 'Do you find the message convincing?' The students' task is to think about the questions as they watch the commercial.
4. Play the commercial.
5. The students work in groups of three or four, discussing the questions on the board and their responses to them.
6. Play the commercial again, for the students to confirm or modify their responses to it.

Homework

The students either rewrite the commercial as they would prefer to see it presented *or* write a commercial for a similar product or service.

TV viewing questionnaire

1 How many hours of television do you watch per day?

  How many hours per week?

2 How many commercials do you think you see per week?

3 To what target audience groups (home owner, pet owner, food shopper, health-conscious consumer, etc.) do you think you belong?

4 At what time of day or night would an advertiser present a commercial aimed at you?

5 With what type of TV programme would an advertiser want to present a commercial aimed at you?

Creative captions

Level Elementary and above

Purpose Viewing comprehension, Written composition

Sequence type Drama

Sequence length 1–3 minutes

Activity time 20–30 minutes

Preparation

Select a suitable sequence and decide what kind of captions you want the students to write (dialogue captions, 'thought bubble' captions or narrative captions). Bring five very large sheets of paper for each group of students, plus one felt-tipped marker per group.

In class

1. Divide the class into small groups and give each group the necessary paper and marker.
2. Tell the students that you are going to play a sequence twice without sound. Their task is to imagine that it is a scene from an old-fashioned silent movie and to write up to five captions to go with the sequence. Indicate what kind of captions they must write. Explain that you will pause occasionally during the second viewing to give them time to write.
3. Play the sequence with the sound turned down. Then play it again, pausing from time to time to allow groups to write their captions on their sheets of paper.
4. Groups take turns going to the front of the room. One student from each group replays the video sequence, pausing at appropriate points to allow other members of the group to hold up their captions by the monitor screen.
5. Play the sequence with the sound turned up. Compare some of the captions with the actual dialogue.

Variation: beginners and above

■ Repeating captions

As a speaking exercise students can watch the sequence and repeat the captions aloud.

Culture comparison

Level Elementary and above

Purpose Cross-cultural comparison, Note-taking, Speaking, Testing, Written composition

Sequence type Documentary, Drama

Sequence length 1–3 minutes

Activity time 15–20 minutes

Preparation

Select a sequence which illustrates several features of the target culture (e.g. UK, USA) which are different from the students' own culture.

In class

1. Ask the students to take a sheet of paper and draw a line down the middle. At the top of the sheet they write on one side of the line the name of the target culture and on the other side of the line the name of their own country.
2. Tell the students that you are going to play a sequence which contains information about the target culture. Their task is to find three things that are different in their country from what is shown in the sequence.
3. Play the sequence. Allow the students to take notes.
4. After viewing, the students complete their notes on their sheets and tell their neighbour what they have discovered.
5. The students report to you. From your knowledge of the two cultures, correct any misconceptions. Introduce cultural points which you think are significant. If necessary, play relevant parts of the sequence again.
6. The students write any other significant cultural differences on their sheets.

Variations

■ Intermediate and above: behaviour study

Select from the sequence two or three incidents of culturally different behaviour to focus on. Ask the students to observe these incidents carefully. Pause after each incident and elicit information about what the people in the video said and did and how this is different from the students' culture. For example: watch how they greet each other.

Play the relevant part of the sequence and elicit from the class:

(a) Where they are
(b) What they said
(c) What they did (e.g. shake hands)
(d) What their relationship was
(e) How they would act differently in the students' mother tongue and culture

■ Intermediate and above: Cross-cultural comparison

Two incidents or sequences can be contrasted to allow the class to compare behaviour in two different situations in the target culture (e.g. greetings at a party and in an office).

■ Advanced: Written composition

The students write an essay comparing and contrasting the situation in the target culture with that of their own.

Test format

Culture comparison can also be used in testing. Select a drama or documentary sequence where characters exhibit culturally significant moods or behaviour. Write five or six words on the board describing the behaviour and number them.

E.g. (1) *sarcastic* (2) *outwardly polite* (3) *impatient*, etc.

Play the sequence and pause at each behaviour point. The students identify the relevant description. Play the sequence again to check students' answers.

Debate the issue

Level Advanced

Purpose Discussion, Note-taking, Speaking

Sequence type Documentary, TV news programme

Sequence length Any length that will fit into the class time available

Activity time 2 class periods

Preparation

Select a sequence which features a controversial issue.

In class

1. Write a motion on the board related to the topic of the video. E.g. for a sequence on ownership of firearms: *Everyone should have the right to possess a gun for self-protection.*
2. Tell the students that you are going to play a sequence related to the motion. As they watch the video, they are to decide how they feel about the motion.
3. Play the sequence.
4. Tell the students that they are now going to participate in a debate. Ask for volunteers to argue 'pro' or 'con'.
5. Select an equal number of students (between two and four) to form two debating teams. Appoint one student from each team to act as captain. Captains will give their presentations first and summarise their team's argument at the end.
6. If there is time, play the sequence again.
7. For homework, have the speakers prepare their arguments. Tell them that they may speak from notes but that they should not read their presentations. Set a 3-minute time limit for each presentation. Ask the rest of the students to prepare between three and five questions each to ask the speakers. Encourage students to use information from the video in their presentations and questions.

In the next class period

1. Have the two teams sit at the front of the room facing the class.
2. Alternate team presentations. Keep to the agreed time limit.
3. After the presentations and summaries, members of the 'audience' question speakers from either team.
4. End the debate with a class vote on which team they found most persuasive.

Design a music video

Level Intermediate and above

Purpose Discussion, Reading, Written composition

Sequence type Music video

Sequence length 3–5 minutes

Activity time 40–50 minutes

Preparation

Select a music video, preferably one which is unlikely to be familiar to the students. Ideally you should have both the music video and an audiocassette of the song. Duplicate copies of the lyrics and the questionnaire on page 46.

In class

1. Distribute the lyrics. The students discuss the mood and meaning of the song.
2. Distribute the questionnaire. Tell the students that they are going to hear the song without seeing any pictures. Their task is to listen to the song and then discuss the questions.
3. Play the audiocassette or the soundtrack of the music video. (Cover the monitor with a cloth, a coat or a large piece of paper.)
4. The students work in groups of three or four, discussing the questions in Part A of the questionnaire.
5. Groups make up a video treatment for the song. (Part B of the questionnaire.)
6. Groups report their ideas for what a video for the song would look like.
7. Play the music video, this time *showing the pictures*.
8. Lead a whole class discussion of the following questions: Is the video what they expected? Did anything in the pictures surprise them? Do the pictures

add to their understanding of the song? Which do they prefer: listening to the song only, or watching the music video?

9. Play the video again, pausing if necessary to extend or clarify points which have arisen in the previous discussion.

Questionnaire: design a music video

Part A: reaction to the song

After listening to the song, discuss the following questions with the members of your group

- What's your reaction to the song? Do you like it? Why or why not?
- How do you feel when you listen to it?
- Do you think the music fits the words? Why or why not?
- What's your opinion of the singer's voice?
- What do you think about the instruments used and how they are played?

Part B: ideas for a video

Together with the people in your group, discuss your ideas for a video based on this song.

- What would your video look like?

Dialogue cards

Level Beginners and above

Purpose Listening, Reading, Testing, Viewing comprehension

Sequence type Drama

Sequence length 1–2 minutes

Activity time 5–10 minutes

Preparation

Select a sequence with clear, simple dialogue. Write each line of the dialogue on a separate index card with the relevant character indicated or named. Make enough cards to give one set of the complete dialogue to each group of three or four students in your class.

In class

1. Group the students and distribute one set of cards to each group.
2. Tell the students that you are going to play a sequence in which the lines on the cards are spoken. Their task is to watch the sequence without referring to the cards and then put the cards in the order in which they occur in the dialogue.
3. Play the sequence.
4. Groups put the cards in order.
5. Play the sequence again, pausing occasionally if necessary. The students check their answers.
6. Groups practise reading the dialogue aloud.
7. Play the sequence again for reinforcement.

Variations

- The students attempt to put the conversation in order *before* viewing the video.

- Instead of separate index cards, write out the whole dialogue in random order. The class must number the lines in the correct order.

> NOTE: *This variation is also a possible testing format.*

- **Intermediate and above: Jumbled dialogue**

- Write out the dialogue without the character names and cut it into strips. Play the sequence with the sound turned down. The students then construct the dialogue from the strips. Play the sequence again with sound so that the students can check their versions.

Dialogue differences

Level Intermediate and above

Purpose Discussion, Reading, Testing, Viewing comprehension

Sequence type Drama

Sequence length 1–2 minutes

Activity time 10–15 minutes

Preparation

Select a sequence in which register is important and distinctive. Prepare a transcript of the video dialogue, along with another dialogue having the same subject matter but exhibiting differences in register (e.g. Dialogue 1 – informal; Dialogue 2 – formal). Make enough copies of the two dialogues for all the students.

In class

1. Distribute the dialogues.
2. Tell the students that you are going to play a sequence without sound. Their task is to watch the sequence and then select the dialogue which corresponds to it.
3. Play the sequence with the sound *turned down*.
4. The students work in pairs, deciding together which dialogue is more appropriate for the sequence.
5. Play the sequence with the sound *turned up*. The students check their answers.
6. (Optional) Use the activity as a starting-point for more intensive work on register.

Variation: Intermediate and advanced

■ Relationships

Select a sequence containing various registers. Write the relationships of the characters up on the board. E.g.:

(a) *Boss to employee*
(b) *Parent to child*
(c) *Friend to friend*

Play the sequence and pause as each register is clearly established. The students must give the letter which describes the relationship of the characters.

NOTE: *This is a good testing format.*

Dialogue fill-ins

Level Elementary and above

Purpose Listening, Speaking, Testing, Written composition

Sequence type Drama

Sequence length Sufficient for 10–15 lines of dialogue

Activity time 20–30 minutes

Preparation

Select a sequence with 10–15 lines of dialogue, featuring two main characters. Prepare and duplicate a transcript of the dialogue, omitting the lines of one of the characters, as on page 53. There should be only one copy for each student.

In class

1. Distribute the transcript. Tell the students that you are going to play a video version of the full dialogue. Their task is to fill in the missing lines *before* viewing the sequence
2. Have the students work in pairs, filling in the lines and practising the dialogues they have created.
3. When all the students have completed the dialogue, volunteer pairs perform their dialogues for the class. Encourage them to 'read and look up' when performing, i.e. they should say their lines while looking directly at their partners. In this way their performances will be more realistic.
4. Play the sequence. The students compare their dialogues with the video original. Replay if necessary.

Variations

■ Elementary and above

Instead of omitting and filling in complete lines of dialogue, the activity can be based on filling in key idioms, phrases, vocabulary or grammatical items.

■ Elementary and above

This version does not need a transcript of the dialogue. Play the sequence and pause before a key word is said (e.g. a preposition). The students write down the missing word. This is a useful test format.

Dialogue fill-in

(from *It's a Wonderful Life*)

MRS BAILEY: Can you give me one good reason why you shouldn't call on Mary?

GEORGE:

MRS BAILEY: Hmmm?

GEORGE:

MRS BAILEY: Well, she's not crazy about him.

GEORGE:

MRS BAILEY: No.

GEORGE:

MRS BAILEY: Well, I've got eyes, haven't I? Why she lights up like a firefly whenever you're around.

GEORGE:

MRS BAILEY: And besides, Sam Wainwright's away in New York, and you're here in Bedford Falls.

GEORGE:

MRS BAILEY: I don't know about war.

Dialogue matching

Level	Elementary and above
Purpose	Viewing comprehension
Sequence type	Drama
Sequence length	1–3 minutes
Activity time	10–15 minutes

Preparation

Select a sequence with three to five speaking parts. If possible the characters should already be familiar to the students. Prepare a list of the names of the characters, along with a separate list of eight to ten lines of dialogue from the sequence, not necessarily in the order in which they are spoken. Make enough copies for all the students.

In class

1. Distribute the lists (or write the information on the board).
2. Pre-teach unfamiliar words and expressions.
3. Tell the students that you are going to play a sequence without sound. Their task is to watch the sequence and then match each line of dialogue with the character who said it.
4. Play the sequence with the sound *turned down*.
5. The students work in pairs and match each line of dialogue with one of the characters on the list.
6. Play the sequence with the sound *turned up*. The students check their answers.
7. Go through the dialogue lines with the class, asking for volunteers to explain how they knew which character had spoken each line.

Variation

Ask the students to match the names with the dialogue extracts *before seeing* the video.

Fill that video gap

Level Intermediate and above

Purpose Discussion, Listening, Oral composition, Speaking, Written composition

Sequence type Drama

Sequence length Two sequences, each about 1 minute

Activity time 10–15 minutes

Preparation

Select a drama with a strong story-line and several separate incidents. Although you are going to use only two short sequences, you must know the whole video extract well.

In class

1. Tell the students that you are going to play two sequences. Their task is to write a story to connect the two.
2. Play a sequence about a minute long.
3. In groups the students discuss the situation and the characters and report to you.
4. Use the Fast Forward button to advance the video for about 2 minutes of screen time at random (*no picture* on the screen).
5. Play the video for another minute.
6. In groups, the students discuss the second sequence.
7. Elicit the situation from them. Get them to compare obvious differences with the earlier sequence (location, characters, attitudes, what's happening, etc.).
8. Each group composes a bridge story to link the two sequences. They may do this either orally or in writing.

9. A spokesperson from each group reads its bridge story. The students discuss the logic of each story (e.g. Does it fit the two contexts?). They may vote on the most convincing story.
10. Rewind the video to the beginning of the first sequence and play right through to the end of the second sequence.
11. The students compare their stories with the actual story on the screen.

Film reviews

Level Advanced

Purpose Discussion, Written composition

Sequence type Complete feature film

Sequence length Entire length of film

Activity time Depends on length of film and lesson time available.

Preparation

Select a feature film which is likely to appeal to your students. Duplicate the questionnaire on page 60 with one copy for each student. Decide how long the film reviews should be.

In class

1. Distribute the questionnaire. Explain that film reviews usually contain the kinds of information asked for in the questionnaire. Go over the questionnaire to make sure that the students understand the type of information requested.
2. Tell the students that you are going to play a film in its entirety. Their task is to write a review of it, using the questionnaire as a guide.
3. Play the film through without stopping, or in sections if the film covers more than one lesson period.
4. The students work in groups of three or four, discussing the items in the questionnaire.
5. For homework, each student writes a review of the film, covering the items in the questionnaire.

In the next class

6. The students work in groups, reading their reviews to each other.

NOTE: *Students may find it helpful in their writing if you provide them with examples of actual film reviews from newspapers and magazines.*

Film review questionnaire

Facts about the film:

1 What is the title of the film?

2 What kind of film is it (psychological drama, comedy, suspense, etc.)?

3 What actors and actresses appear in the film? What roles do they play?

4 Who directed the film?

5 Who wrote the screenplay?

6 What is the story about? (Do not attempt to summarise the entire film in detail. Just give a general idea of the story, without telling how it ends.)

Your opinions. What do you think about:

7 The acting?

8 The direction?

9 The screenplay?

10 The music or any other elements you consider important?

Your recommendations to other film-goers:

11 Do you recommend that people see this film? Why or why not?

12 What type of film-goer is most likely to enjoy this film?

Find the idea

Level Intermediate and above

Purpose Discussion, Viewing comprehension

Sequence type Music video

Sequence length 3 minutes

Activity time 30 minutes or more

Preparation

Select a music video with interesting or unusual images, or one inspired or influenced by other art or cultural objects. Obtain reproductions, film stills, magazines or books which illustrate the influences or the originals of the images in the video. (E.g. *Hourglass*, by the group Squeeze, was inspired by paintings by Dali and Magritte.) Decide how much pre-viewing explanation the students will need.

In class

1. Tell the students that you are going to show them a music video. Their task is to work out where its influence or inspiration came from.
2. Play the video.
3. The students discuss the influence or inspiration for the video.
4. Show the realia you have prepared.
5. The students discuss the realia and say how these relate to the video. Replay parts of the video as necessary to aid discussion.
6. The students discuss: Was the inspiration appropriate for the song? Could it have been done better?
7. Play the video a final time.

Variation: Intermediate and above

■ Pop interpretation

The students watch the video and study the lyrics. How are the lyrics represented in vision?

Five W's and H

Level Elementary and above

Purpose Active viewing, Note-taking, Speaking, Viewing comprehension, Written composition

Sequence type TV news programme

Sequence length 1–2 minutes

Activity time 10–15 minutes

Preparation

From a TV news programme, select one or more items which have a single news presenter and sufficient content to make full use of the five W's and H.

In class

1. Explain that news programmes usually present factual information concerning the 'five W's and H' of a subject:

    Who is it about?
    What is it about?
    When did it happen?
    Where did it happen?
    Why did it happen?
    How did it happen?

2. Write the six question words on the board. Tell the students that you are going to play a sequence with one or more news items (specify the actual number). Their task is first to watch, then to watch again and make notes.
3. Play the sequence twice. First, the students watch. The second time, they make brief notes under the six headings.
4. The students report back on how much they understood, using the six headings.
5. Play the sequence a third time for reinforcement.
6. As a follow-up, students can write a short news item based on the sequence.

From screenplay to film

Level Intermediate and above

Purpose Discussion, Reading, Viewing comprehension

Sequence type Complete feature film

Sequence length Depends on length of film

Activity time Between eight and ten 15-minute 'reading sessions' plus a final session (or sessions) long enough to show the entire film, without interruption if possible

Preparation

Select a feature film which is likely to appeal to the students concerned and for which a complete screenplay is available. Prepare enough copies of the screenplay to give one to each student, or make arrangements to ensure that each student can read the entire screenplay before viewing the film.

Prepare your reading sessions carefully. First, check on any differences between the published text and the video which you will show to the class. Then decide exactly how you are going to 'read' the screenplay aloud. How are you going to deal with the different characters? How are you going to describe the action (i.e. are you going to read the descriptions given in the screenplay, or make up your own description)? What items are you going to explain to the class? Are you going to describe parts of the film without attempting to read the dialogue?

In class

1. Distribute the screenplay. Tell the students that they are going to see a feature film after becoming familiar with the screenplay. Explain that you will read scenes from the screenplay to them in each class period over the next few weeks. Their task is simply to follow and concentrate on understanding the story. If necessary, warn them about differences between the text of the screenplay and the actual readings you will give.

2. At the end of each period, read part of the screenplay to the class. As you read, stop from time to time to explain key items, and the cultural context in

which they appear, for students who may not understand, and allow them to ask questions. If you allot 10–15 minutes per period for these 'reading sessions', it should take roughly eight to ten periods to get through an entire screenplay. Some students will go ahead and read the entire screenplay on their own. Do not discourage this.

3. After you have finished reading the screenplay, show the entire film, possibly during the last class meeting of the year or course. Do not give the students any worksheets or activities in connection with the film. They are simply to watch and enjoy it.

NOTE: *Large book shops and others specialising in film and drama, stock screenplays, as well as dramas which have been converted to film. Note that screenplays often differ quite substantially from the video versions of feature films available to you and your students. These versions are often shortened and edited in various ways.*

Guess the picture

Level Elementary and above

Purpose Viewing comprehension

Sequence type Any

Sequence length 30 seconds to 1 minute

Activity time 15 minutes

Preparation

Prepare two templates of cardboard, one which is square and covers the centre of the screen leaving a margin of about 20 cms around the edge of the screen. The other is square and big enough to cover the screen, with a circular hole in the middle about 30 cms in diameter. Cut each template in half, allowing you to make the space larger or smaller as necessary.

In class

1. Explain to the students that they are going to watch a sequence with only part of the picture revealed. Their task is to work out the content of the whole picture and the story or situation.
2. Use one of the templates to cover either the centre of the screen or to reveal a hole in the centre.
3. Play the sequence with sound and vision.
4. The students make a first attempt at guessing the picture and the story or situation.
5. Replay the sequence as many times as necessary. Gradually enlarge or reduce the template to reveal more of the picture each time.

How much can you see?

Level Intermediate and above

Purpose Discussion, Speaking, Written composition

Sequence type Any

Sequence length 1–5 minutes

Activity time 15–20 minutes

Preparation

Select a sequence which will effectively exploit the technique below. Drama and documentary events work best. Check out the effectiveness of the sequence in the classroom. Re-position the monitor screen if necessary.

In class

1. Get half the class sitting in profile to the monitor screen facing one wall, and the other half in profile facing the opposite wall. Ask the students to concentrate on something on the wall.
2. Tell the students that you will play a sequence and they will see it out of the corner of their eyes. At the end of the sequence, they must reconstruct it as best they can. To prevent them sneaking a look, present the task as a challenge: 'How much can you see out of the corner of your eye?'
3. Play the sequence with sound and vision.
4. The students work in groups of four to reconstruct the sequence. Each group should have two pairs who were facing opposite walls.
5. Each group writes the story of the sequence in not more than ten sentences.
6. A spokesperson from each group reads out the story which is then discussed.
7. Replay the sequence. The students watch and compare the reality with their perception when using peripheral vision.

In character

Level Intermediate and above

Purpose Note-taking, Speaking, Written composition

Sequence type Drama

Sequence length 2–5 minutes

Activity time 15–20 minutes

Preparation

Select a sequence with a strong story-line involving at least two characters.

In class

1. Tell the students that you are going to play a sequence twice. Their task is to write a first-person, past-tense summary of what happens from the point of view of one of the characters. During the first viewing they simply watch and get a general idea of what happens. During the second viewing they may take brief notes on the action and dialogue. They should choose their character before the second viewing.
2. Play the video sequence *twice*. If necessary, pause occasionally during the second viewing to allow the students to take notes.
3. The students work individually, writing their summaries in the past tense. To get them started, give an example on the board, e.g.:

   The train stopped and my brother Harry got off. I ran to him and we embraced.

4. When students have finished writing, ask them to work in small groups, taking turns to read their summaries to the rest of their group.
5. Play the sequence again for the students to check the accuracy and appropriateness of their summaries.

In other words

Level Intermediate and above

Purpose Grammar review, Listening

Sequence type Music Video

Sequence length 3–4 minutes

Activity time 20–30 minutes

Preparation

Select a music video with suitable examples of non-standard forms. Decide how much help the students will need to carry out the activity, depending on the language level and difficulty of the lyrics.

In class

1. Explain to the students that pop music lyrics often contain non-standard forms which may not be appropriate in more formal, written English. Write an example or two on the board (e.g. 'My man *ain't took* me *nowhere*', 'I'm *gonna* love you', etc.) and ask for volunteers to go to the board and write the equivalent in standard, written English.
2. Tell the students that you are going to play the music video *twice*. Their task during the first viewing is to watch and try to spot non-standard forms, without writing anything down. During and after the second viewing they are to write down the non-standard forms they hear.
3. Play the music video twice. If necessary, pause from time to time during the second viewing.
4. The students work in groups of three or four, comparing their notes and writing a standard English equivalent of each non-standard form they heard.
5. Volunteers go to the board and each writes one non-standard item he or she heard, along with a standard English equivalent.

Variation

Provide students with a list of examples of non-standard English from the song and ask them to write the equivalents *before* they watch the music video.

In the mood

Level Elementary and above

Purpose Discussion, Viewing comprehension, Vocabulary development

Sequence type Drama

Sequence length 1–3 minutes

Activity time 15–20 minutes

Preparation

Select a sequence with two or three characters whose dialogue and behaviour display different moods.

In class

1. Write a list of five or seven adjectives describing different moods on the board, e.g.:

interested	*happy*	*bored*	*frightened*
uninterested	*sad*	*angry*	*indifferent*

   Make sure that the students understand the words listed and that you have included the moods displayed by the characters.

2. Tell the students that you are going to play a drama sequence in which each of the characters is in a particular mood indicated by one of the words on the board. Their first task is to identify the mood of one character. Describe the character concerned.
3. Play the sequence.
4. Call on individual students and ask them to identify the mood displayed by the character (e.g. 'How does Charlie feel?').
5. Play the sequence again, this time pausing after each line spoken by the chosen character. Call on individual students to provide a line of dialogue indicating a *different mood* from that shown in the video (e.g. 'How would Charlie say that line if he were angry?').

6. Extend the activity to include the other characters and their particular moods, but always concentrate on one character at a time.

Variation

■ Follow the change of mood

A character's mood may change within a sequence (e.g. if he or she hears some particularly bad or good news). On the board put the beginning and end mood of a character. E.g.:

Happy ______________ *Sad*

Play the sequence, pausing at times if necessary. The students identify the changes in mood, and the behaviour and words which illustrate those changes.

Instant interview

Level Intermediate and above

Purpose Speaking, Viewing comprehension

Sequence type Any

Sequence length Any

Activity time 10–15 minutes

Preparation

Select a suitable sequence, possibly one which the class has already studied and found difficult. (See *NOTE* below.)

In class

1. Pre-teach these interview questions:

   Excuse me. I'm from the (town or school name) Post newspaper.
   What's happening?
   What did they say?
   What's your opinion?
   or
   What do you think will happen?

2. Tell the students what kind of sequence you are going to play. Explain that while the video is playing, any student can be a reporter from the (X) Post newspaper. All they have to do is get up, stop the video (using the Pause button) and ask another student one of the questions above.

3. Play the sequence.

4. As you play the sequence, pause a couple of times and ask the questions yourself. If the 'interviewee' can't answer, ask another student to play that part of the sequence again and repeat the question.

5. Encourage one or two students to get up and conduct instant interviews. Soon everyone will be doing it.

> NOTE: *This is an excellent way of encouraging students who haven't understood the video to get help. It is also an excellent way of reviewing a sequence the class has studied. For a more formal interview, see 'Opinion poll'.*

Interior monologue

Level Intermediate and above

Purpose Speaking, Viewing comprehension, Written composition

Sequence type Drama

Sequence length 1–3 minutes

Activity time 30–45 minutes

Preparation

Select a sequence in which one character clearly has strong feelings or reactions but speaks little or no dialogue.

In class

1. Tell the students that you are going to play a drama sequence in which one character says very little but obviously has thoughts and feelings about what is happening. Their task is to imagine what that character is thinking and feeling and to write an interior monologue from his or her point of view.
2. Play the sequence, twice if necessary.
3. The students work in groups of three or four, discussing what the character is thinking or feeling. Then they write the interior monologue required.
4. Each group reads its monologue to the class.
5. Play the sequence again, pausing and discussing points of detail if necessary.

Juke box jury

Level Intermediate and above

Purpose Discussion, Note-taking

Sequence type Music video

Sequence length 3 minutes

Activity time 30 minutes

Preparation

Select a music video which is likely to arouse discussion, and about which students are likely to hold differing opinions.

In class

1. Write the following chart on the board:

Juke Box Jury Chart (score out of 10 points)			
Tune	*Lyrics*	*Performance*	*Presentation*

2. Tell the students that you are going to play a music video. Their task is to judge it and award points according to the chart.
3. Play the video twice, allowing the students time to make notes between the two viewings.
4. Divide the class into three groups. Each group discusses the video and appoints a spokesperson.
5. Each spokesperson gives his or her group's scores under the various headings of the chart. The groups' scores are compared and discussed.
6. Play the video again.
7. A final class vote is taken. If the video is new, will the song and/or the video be a hit or not? If it is not new, was its success or failure deserved?

Variation

The same activity can be extended to include two or more music videos.

Listen and say

Level Beginners and above

Purpose Listening, Pronunciation

Sequence type Drama

Sequence length 1–2 minutes

Activity time 5–10 minutes

Preparation

Select a sequence with fairly simple dialogue and no overlapping speeches.

In class

1. Tell the students that you are going to play a sequence twice. During the first viewing their task is simply to follow the story. Explain that during the second viewing you will stop the video from time to time so that they can repeat the line just spoken.
2. Play the sequence without stopping.
3. Play the sequence again, pausing to single out lines for choral repetition. Encourage the students to use the same intonation as the character.

Variation

During the second viewing, rather than have the class as a whole repeat each line, ask individual students to repeat the line.

News summary

Level	Elementary and above
Purpose	Speaking, Viewing comprehension
Sequence type	TV news programme
Sequence length	5–15 minutes
Activity time	15–45 minutes

Preparation

Select all or part of a news programme, with several different stories. Duplicate the chart opposite, with a copy for each student.

In class

1. Distribute copies of the chart. Make sure the students understand the headings listed.
2. Tell the students that you are going to play part of a TV news programme. Their task is to complete the chart by noting the most appropriate heading for each of the stories.
3. Play the sequence
4. The students work individually, completing the chart.
5. Replay the sequence, pausing or stopping the video after each story. Volunteers say which heading they chose, and the class discusses the appropriateness of the choice if there is any disagreement.

News summary

Foreign news						
National news						
Local news						
Accidents						
Disasters						
Violence						
Famous people						
Arts/entertainment						
Sports						

Opinion poll

Level Intermediate and above

Purpose Discussion, Speaking

Sequence type Documentary, Interview/Talk show, TV news programme.

Sequence length 5–10 minutes

Activity time 20–30 minutes

Preparation

Select a sequence which is likely to arouse strong and, if possible, opposing opinions.

In class

1. Put the lists below (or something similar, to suit the sequence selected) on the board. Pre-teach different ways of stating an opinion, using the lists.

+ *I loved it.*	– *I hated it.*
I liked it.	*I didn't like it.*
I enjoyed it.	*I was upset by it.*
I was delighted by it.	*I was horrified by it.*
I thought it was interesting.	*I thought it was boring.*

2. Tell the students that you are going to play a sequence about which they are sure to have an opinion. Their task is to watch the sequence and then interview each other to find out the different opinions held by members of the class.

3. Play the sequence.

4. In pairs the students ask each other, 'What did you think of the programme?' or 'How did you like the programme?' and give reasons for their answers.

5. After the pair work, ask the whole class, 'Who loved it?', 'Who hated it?', etc. The students answer by raising their hands to indicate their opinions. Tally up the number of students who indicate each of the different opinions.

6. If you wish, the students can draw up a new chart with a survey of the different opinions.

NOTE: *For a more informal activity for stating opinions, see 'Instant interview'.*

People in the news

Level Intermediate and above

Purpose Listening, Note-taking, Speaking, Written composition

Sequence type TV news programme

Sequence length 2–5 minutes

Activity time 15–20 minutes

Preparation

Select a sequence with several items about famous people. Decide how many times you will need to play the sequence to make the activity work with your class. Make copies of the chart opposite for all the students.

In class

1. Distribute the chart. Tell the students that you are going to play a TV news sequence containing a number of items about famous people. Their task is to listen to the ways different people are identified and referred to and then fill in the chart.
2. Write an example on the board, e.g.:

George Bush	*He*	*President Bush*
Mr Bush	*The President*	*The Commander-in-Chief*

3. Play the sequence more than once if necessary.
4. The students work individually, filling in the chart.
5. Volunteers go to the board and each one writes the name of one person and the different ways he or she was referred to. The students check their answers.
6. Play the sequence again, pausing at the end of each item for a final check.

Variation: Advanced

Using the information from the news programme and possibly additional library research, the students in groups prepare a written profile of one of the people referred to.

People in the news

Name of person	Referred to as . . .

Predict the plot

Level Intermediate and above

Purpose Active viewing, Oral composition, Written composition

Sequence type Drama

Sequence length 3–5 minutes

Activity time 20 minutes

Preparation

Having selected a suitable sequence, choose up to ten words or phrases which give a clue to the story, the way characters behave, etc. These items should be genuinely helpful for writing a narrative synopsis of the sequence.

In class

1. Write the chosen words and phrases on the board in programme order. Make sure the students understand the meaning of each one.
2. The students brainstorm story possibilities incorporating the items on the board.
3. In pairs, the students write a plot synopsis. While they are writing, go round, looking at synopses and giving help where necessary.
4. Select some students to read out synopses representing different story possibilities.
5. Play the sequence.
6. The students compare their synopses with the plot of the sequence. Which synopsis was closest to the video?
7. Play the sequence again, as a final check.

Variation

As an addition to stage 5 above, stop before the end of the sequence. The students write a paragraph saying how the sequence will end, based on what they have just seen.

Prediction

Level	Beginners and above, children
Purpose	Active viewing, Discussion, Speaking, Vocabulary development
Sequence type	Any
Sequence length	1–5 minutes
Activity time	30 minutes

Preparation

Select a sequence from a TV news programme, drama or feature film, in which the situation is quickly and clearly established. Prepare a list of the following headings: TOPIC, SIGHTS, WORDS, SOUNDS, SMELLS. Make copies for all the students.

In class

1. Distribute copies of the list (or write it on the board). Tell the students that you are going to play the beginning of the sequence. Their task is to predict what the whole sequence will be like in terms of the headings in the list.

2. Play enough of the sequence to establish the topic.

3. In pairs the students discuss and write down the following under each heading.

TOPIC	the subject of the sequence
SIGHTS	things they expect to see
WORDS	words they expect to hear
SOUNDS	sounds they expect to hear
SMELLS	things they might smell if they were there

4. Elicit from the students some of their ideas.

5. Play the rest of the sequence.

6. The students in pairs discuss what sights, words, sounds, etc., occurred in the sequence and where.
7. The pairs report back on results.
8. Play the sequence again, with more intensive study if required.

Variations

■ Elementary and above

As part of the preparation, decide on five or six 'pause' points in the sequence. In class, pause or stop at each point. The students predict what will happen next.

■ Beginners

Tell the class the subject of the sequence, e.g. a murder. The students predict the answers to the following questions:

Who will you see?
Where will the scene or action take place?
What things will you see?
What are some of the lines of dialogue you will hear?

Depending on the subject-matter, you may be able to ask the students these questions before playing the sequence, or you may need to play the beginning first.

■ Elementary and above

Play the sequence and pause at the end of each line of dialogue. The students predict the next line.

> TIP: *Instead of releasing the Pause button, press Stop and then Play. On most video recorders the videocassette rewinds a few seconds before starting to play again. This allows the class to hear the previous line again.*

■ Intermediate and above

Make an audio recording of the soundtrack and divide the students into two groups. One group watches the video without sound and predicts the topic, words and sounds. The other group listens to the audiocassette and predicts the topic, sights and smells. You will need two rooms for this variation.

■ Children: beginners and above

Tell the children the subject of the sequence and write down six objects or people, four of which appear in the sequence and two of which do not. The children predict which ones will appear.

> NOTE: *Active viewing worksheets involving prediction can be an excellent way of pre-teaching essential vocabulary.*

Read and say

Level Beginners and above

Purpose Listening, Reading, Speaking

Sequence type Drama

Sequence length 1–3 minutes

Activity time 10–20 minutes

Preparation

Select a short sequence with simple dialogue and as many characters as possible. Prepare a transcript of the dialogue and make a copy for each student.

In class

1. Distribute the transcript.
2. Divide the class into groups. Assign a character to each group.
3. Tell the students that you are going to play the sequence in which the dialogue occurs. Their task is to listen to how the dialogue is spoken, and later take the part of the character they have been assigned.
4. Play the sequence two or three times.
5. The students read the dialogue in chorus, with the various groups reading the parts they have been assigned.
6. After two or three readings, the groups switch characters and read the newly assigned parts. Play the sequence as many times as necessary to help the students' performance.

Variation

In a sequence containing captions with model phrases, the students can repeat the phrases as they appear on the screen.

Reconstruction

Level Intermediate and above

Purpose Discussion, Note-taking, Reading, Speaking, Viewing comprehension

Sequence type Drama

Sequence length 1–2 minutes

Activity time 15–20 minutes

Preparation

Select a sequence with a clear, visual narrative line. Sequences from silent films work particularly well.

In class

1. Explain to the students that you are going to play a sequence in which a particular incident (e.g. a robbery) takes place. Their task is to observe details, and then describe what they have seen in chronological order, listing as many details as possible.
2. Play the sequence.
3. The students work in groups of three or four, discussing what they have seen and preparing a list of events in the correct order.
4. A volunteer from each group reads its list to the class.
5. The class vote on which list was the most accurate and detailed.
6. Play the sequence again. Pause from time to time to allow volunteers to report on what they have just observed happening on the screen.

Variation

While replaying the sequence, pause and ask students to report on what is going to happen just after the still frame on the screen.

Roleplay

Level Elementary and above

Purpose Listening, Speaking

Sequence type Drama

Sequence length 1–3 minutes

Activity time 20–30 minutes

Preparation

Select a sequence with two or more characters and clear dialogue.

In class

1. Tell the students that you are going to play a sequence twice. Their task is to study the situation in the video, and then roleplay the same situation using whatever words or other means they wish.
2. Play the sequence twice.
3. Divide the class into groups composed of the same number of students as there are characters in the sequence. Allow the groups 5 minutes or so to rehearse roleplaying the situation in the video.
4. The groups take turns performing the situation for the class, using their own words, actions and gestures.
5. Play the sequence again, and compare it with the roleplays.

Variation

In large classes each group prepares the role of *one* character. It then nominates one of its members to play that character. During the roleplay, players can be substituted by the teacher calling 'Change!' or by another group member tapping the roleplayer on the shoulder and replacing him or her.

Scene study

Level Intermediate and above

Purpose Reading, Speaking

Sequence type Drama

Sequence length 1–2 minutes

Activity time 30–45 minutes

Preparation

Select a sequence with two speaking roles. If possible, allow for the variation below. Prepare a transcript of the dialogue and also a one-sentence description of the situation (e.g. There is a meeting between a man and a woman who have not seen each other for 20 years). Make a copy for each student.

In class

1. Distribute the transcript.
2. Tell the students that you are going to play a video version of the transcript. Before this, however, they must read the transcript, think about the characters' intentions and feelings, choose one of the roles, and then rehearse the dialogue with a partner as they think it *might be played* in the video.
3. While the students are rehearsing, circulate among them, answering questions and providing help as needed.
4. Pairs take turns giving their dramatic interpretations of the scene for the class.
5. Play the sequence. Students look out for similarities and differences between their own interpretations and the video version.

Variation

Have the students work in groups of three. Ask them to add a new character to the scene and to act out what the scene would be like with the new character in it.

See the film, write the book

Level Intermediate and above

Purpose Oral composition, Speaking, Written composition

Sequence type Drama

Sequence length 5 minutes or more

Activity time 30 minutes

Preparation

Select a drama sequence with plenty of action and/or dialogue, and no voice-over narration or commentary.

In class

1. Play the sequence. The students study it for subject-matter and language.
2. Explain to the students that many TV series and films are accompanied by books which turn the series or film into a novel. Their task is to write a novelisation of the sequence they have studied.
3. Brainstorm with the students key points that need to be included and write these on the board.
4. Elicit (or teach if necessary) some appropriate descriptive words and expressions.
5. The students write the first paragraph of the novelisation, then read it out. Suggest improvements, etc.

Homework

The students write their novelisations in 100 words or more.

In the next class

Correct the novelisations. Read out or put on display the most 'publishable'.

Variation

If the sequence is long, different groups can be 'commissioned' to novelise different 'chapters' of the story in 50–100 words. At the end all the 'chapters' are read out in sequence to produce one long 'novelisation'.

Sequencing

Level Elementary and above

Purpose Discussion, Oral composition, Reading, Viewing comprehension

Sequence type Documentary, Drama

Sequence length 3 minutes or longer

Activity time 10 minutes

Preparation

Select a sequence in which the order of events is important or complicated. Write not more than ten sentences summarising the events in the sequence. Put them in random order and number them. Make a copy for each student.

In class

1. Tell the students that you are going to play a sequence, and tell them how long it is. Their task is to take particular notice of the order of events.
2. Play the sequence.
3. Distribute copies of the sentences.
4. In pairs, the students put the sentences in the same order as the events in the video sequence.
5. Elicit the results from the students. If there is disagreement on the order, play the relevant part of the sequence again to check.
6. Play the whole sequence again for reinforcement.

Variations

■ Ordering

Instead of sentences summarising the events, choose not more than ten content words that occur in the sequence and put them in random order. The students must then put them in the correct order. If the level of language allows, the students can then reconstruct the whole sequence orally.

■ Intermediate and above

To provoke discussion, include an item in the summary that could go in two different places in the order of events. The students discuss the correct position for the item.

NOTE: *Sequencing is an excellent way of ensuring that all the students agree on the content of the video sequence before proceeding to other activities arising from it.*

Sex change

Level Intermediate and above

Purpose Cultural awareness, Discussion, Written composition

Sequence type Drama

Sequence length 3–5 minutes

Activity time 20–30 minutes

Preparation

Select a sequence with both male and female characters, which has interesting potential in terms of this activity.

In class

1. Tell the class that you are going to play a sequence in which there are both male and female characters. Their task is to imagine how the sequence would differ if each character were the opposite sex to that portrayed in the video.
2. Write the following questions on the board:

   If each character were the opposite sex, what differences would there be in:

   - *The language of the characters?*
   - *Their clothing?*
   - *Their behaviour?*
   - *The story as a whole?*
3. Play the sequence.
4. The students work in groups of three or four discussing the questions.
5. One student from each group reports to the whole class.
6. Play the sequence again, pausing if necessary to focus on points brought up by the students.

Variation

■ Written composition

As an extension of the activity the groups can rewrite the dialogue to illustrate the differences they have discussed and any problems they have encountered.

Silent viewing

Level Beginners and above

Purpose Active viewing/speaking

Sequence type Any

Sequence length 30 seconds to 1 minute

Activity time 10–15 minutes

Preparation

Select a sequence with clearly identifiable location, characters and situation.

In class

1. Pre-teach essential vocabulary (only if necessary).
2. Brief the students on the situation (only if necessary).
3. Give the students the following questions:

   Where are the people?
   Who are they?
   What's happening?

4. Play the sequence with the sound *turned down*.
5. Elicit answers to the questions. If possible, extend to include discussion of the answers.
6. Play the sequence again with sound and vision. The students compare their guesses with the actual content of the video.
7. (Optional) Proceed to intensive study of the sequence for content and language.

Variation

■ Beginners

Write on the board descriptions of three characters, locations and situations, with *only one of each* corresponding to the video. The students copy all of them down. Teach any new words, then play the sequence. The students circle the character, location and situation they see in the video.

■ Intermediate and above

Increased ambiguity of location, characters and situation are acceptable at this level. Starting at the beginning, reveal the sequence bit by bit, using the pause button. The students guess the location, characters and situation. Opinions will change as more of the sequence is gradually revealed.

NOTE: *Silent viewing is a basic technique using video for three reasons:*

1. *It encourages class communication right from the start.*
2. *It splits aural and visual stimuli, thus allowing the student to form an impression of the situation and likely language used before concentration on the language itself.*
3. *It encourages concentration on paralinguistic features for comprehension.*

Sound only

Level Elementary and above

Purpose Listening, Speaking

Sequence type Drama, Documentary

Sequence length 30 seconds to 2 minutes

Activity time 15 minutes

Preparation

Select a sequence with a sound-track suitable for the activity described below. Prepare a list of three to five questions about the sequence (e.g. Where does the scene take place? How many characters are in it? How are they dressed? What are their ages? What is their relationship? What are they doing?).

In class

1. Distribute copies of the list, or write it on the board.
2. Tell the students that they are going to hear the sound-track of a sequence *without seeing the pictures*. Their task is to listen to the words, sound effects, music, etc., and predict what they will see in the pictures.
3. Play the sequence with *sound only*. (Cover the monitor screen with a cloth, a coat or a large piece of paper.)
4. The students work in groups of three or four, discussing the questions and giving reasons for their answers.
5. Play the sequence again, this time *with sound and vision*.
6. Groups discuss the questions again, taking into consideration the new information they have from the pictures.
7. Play the sequence again for reinforcement.

Variations

■ Elementary

Make a list of adjectives describing the characters in the sequence. Pre-teach them if necessary. E.g.:

nice	*nasty*
tall	*short*
dark	*fair*

Play the sequence with sound only. The students attribute characteristics to the voices they hear. Play the sequence again, with sound and vision. The students compare what they expected with what they actually see in the video.

■ Elementary and above

The students draw the people they hear, or what they hear described or referred to. They describe their drawings to their neighbours. Ask some students to describe their drawings to the class. The drawings will contain personal idiosyncrasies which you can ask about; e.g.:

A restaurant inspector — Bald
Question: Why is he bald?

Encourage students to ask about such unusual aspects of each other's drawings. Finally, play the sequence with sound and vision and discuss the drawings in the light of what the video reveals.

Sound search

Level Elementary and above

Purpose Pronunciation

Sequence type Any

Sequence length 1–3 minutes

Activity time 10–15 minutes

Preparation

Select a sequence with clearly spoken dialogue, commentary, etc. Prepare a transcript. Make enough copies for all the students.

In class

1. Pre-teach the sound you wish to focus on, e.g. [æ]. Ask the class for examples of words containing the sound. Write the examples on the board.
2. Distribute copies of the transcript. Tell the students that you are going to play a sequence in which there are several words containing the sound. Their task is to listen while reading the transcript and to circle all the words containing the sound.
3. Play the sequence
4. The students work in pairs, comparing answers.
5. Play the sequence again. Pause each time the sound occurs. Ask for volunteers to say the word containing the sound.

Speech bubbles

Level	Elementary and above
Purpose	Speaking, Viewing comprehension
Sequence type	Drama, Interview
Sequence length	30 seconds to 3 minutes
Activity time	15 minutes

Preparation

Select a sequence featuring two or three speaking parts. Prepare a speech or thought bubble out of cardboard, large enough to be seen by everyone in the class. It should look like this:

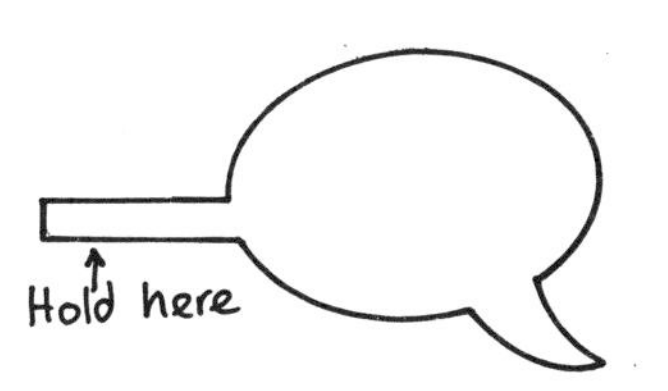

In class

1. Tell the students that you are going to play a sequence twice without sound. Their task is to imagine what the characters or participants are saying.
2. Play the sequence through without stopping and with the sound *turned down*.
3. Tell the students that you are going to replay the sequence, pausing each time someone on the screen speaks.
4. Replay the sequence. Pause each time someone on the screen speaks. Hold the speech bubble over the speaker's head and invite volunteers to say what they think the character is saying.
5. Play the sequence *with sound*. Students compare their guesses with the actual content of the conversation.

Variations

- The students view the sequence the first time *with sound*. Turn the sound down for the second viewing. Each time you pause, individual students attempt to *recall* the exact words used by the character/speaker at that point in the sequence.
- Instead of imagining or recalling what the character/speaker is saying, students say what the character/speaker is *thinking*. This variation works best with sequences containing reaction shots.
- Once the students get used to the activity, allow them to operate the pause button and to use the bubble themselves.

Sports watch

Level Elementary and above

Purpose Discussion, Listening, Note-taking, Written composition

Sequence type TV sports summary

Sequence length 2–5 minutes

Activity time 10–15 minutes

Preparation

Select a sequence containing enough detail to exploit the chart on page 106. Duplicate the chart, with a copy for each student.

In class

1. Distribute the chart and make sure the students understand the type of information required to complete it.
2. Tell the students that you are going to play a TV sports summary. Their task is to complete the chart with information from the summary.
3. Play the sequence once (more times if necessary, depending on its complexity and on the language level of the students).
4. The students work individually, completing the chart.
5. When they have finished, the students compare each others' answers.
6. Play the sequence again, pausing from time to time as the students check their answers.

Variation: Advanced writing

Using the charts, the students in groups write their own sports summaries and then 'broadcast' them. Their performances can be recorded with a video camera and then played back for intensive study.

Sports information chart

What sport?	Where?	Who competed?	Who won?	What was the score or time?

Stills

Level Elementary and above

Purpose Speaking, Vocabulary development

Sequence type Documentary, Drama

Sequence length 30 seconds

Activity time 10 minutes

Preparation

Choose a short sequence which has a background or location with lots of visual detail that can be used like a wallchart. Select a lexical set that can be taught from the sequence. (E.g. for a street scene: items of transport.)

In class

1. Play the sequence, and pause at a convenient point to give a still picture.
2. Teach words from the picture.
3. Present the words in sentences on the board.
4. Use the same still picture for reinforcement/review. The students say what they see.
5. Play the complete sequence, to put the still picture into context.

Variation

Instead of using a single still picture, pause at several different points, where specific items to be taught come into prominence.

Stop/start viewing

Level Elementary and above

Purpose Grammar presentation, Grammar review, Speaking, Viewing comprehension, Vocabulary development, Vocabulary review

Sequence type Any

Sequence length 30 seconds to 5 minutes

Activity time 10–30 minutes

Preparation

Most language teaching video, and much authentic video, can provide suitable sequences for this basic activity, which can be integrated with other tasks and activities and which you can select by consulting the Index of Activities (p. 170). Identify the points of comprehension and language items you wish to focus on. Decide on the board presentation of selected language items.

In class

1. Tell the students what kind of sequence you are going to play, and give them an Active-viewing task and possibly a Sequencing task.
2. Play the sequence, pausing at selected points for comprehension and/or language focus. At each point elicit detailed comprehension of the situation. Then elicit or teach the chosen language item. Get students to repeat the item in chorus or individually.
3. At the end of the sequence, use the board to summarise the main points you wish the students to remember.
4. (Optional) The students do back-up textbook exercises or drills on the language points presented.
5. Play the sequence through for reinforcement. Possibly do a Cross-cultural comparison activity.
6. (Optional) Follow up with reading, writing, discussion or roleplay activities.

Variation: Intermediate and above

With a small class, or with a group in a situation where the rest of the class is doing another activity, the students can initiate their own Stop/Start study. They watch the sequence in its entirety, then replay it, pausing after each sentence to discuss the meaning of new words and phrases. They note anything unclear for discussion later.

Synonym match

Level	Elementary and above
Purpose	Vocabulary review, Vocabulary development, Written composition
Sequence type	TV commercial
Sequence length	30–60 seconds
Activity time	5 minutes

Preparation

Select a commercial which depends on a few key words or expressions. Prepare a list of synonyms for key items occurring in the dialogue or commentary of the commercial, and make copies for all the students.

In class

1. Distribute the list of synonyms.
2. Tell the students that you are going to play a TV commercial in which a synonym for each word or expression on the list is used. Their task is to listen for the synonyms and match them with the items on the list.
3. Play the commercial as often as necessary.
4. The students match the items, then exchange and compare their answers.
5. Play the commercial again. Pause after each key word or expression is used. Ask for volunteers to say which synonym on the list goes with the word or expression just spoken.

Variation

■ Written composition

The students write scripts for commercials for another product or service, using key words or expressions from the commercial viewed.

Team trivia quiz

Level Beginners and above, Children

Purpose Discussion, Grammar review (question forms), Reading, Speaking, Viewing comprehension

Sequence type Any

Sequence length Up to 15 minutes

Activity time Up to 30 minutes

Preparation

Select a sequence which includes a lot of details that can trigger the kind of questions referred to below.

In class

1. Divide the class into two teams.
2. Tell the students the general nature of the sequence you are going to play. The task of each team is to draw up a list of questions about visual details in the sequence. Give examples of the kind of questions you expect: e.g. 'What time did the church clock say?', 'What colour was the man's shirt?', 'Was the woman wearing high heeled or flat shoes?', etc.
3. Play the sequence.
4. In pairs, each team writes down questions. Go round helping with vocabulary. Allow 5 minutes for this.
5. When the teams have drawn up their questions, the quiz begins. Each team asks the other team a question in turn. The team can specify which student must answer. Scoring is as follows:

    1 point per correct answer
    1 point per correct expression
    1 point deducted for asking a question twice

The team with the most points wins.

6. If necessary, replay parts of the sequence to confirm questions and answers. In any case, you should play the entire sequence again at some point.

Variations

■ Beginners

Set the questions and study them with the students *before* they watch the sequence.

■ Elementary and above

Prepare a list of 15–20 questions requiring the students to make specific observations about the sequence, e.g. 'What colour dress does Jane wear to the party?', 'Who drops a pen?', etc. The students must find answers as they view. After viewing, the students, working in pairs, complete the quiz. Check to see who gets it all right on the first viewing.

■ Children: Viewquiz

Prepare a list of five or six questions which summarise the story of the video. In class, give the children the questions. Play the sequence, at first studying it for both content and language. After a second or third viewing, the children answer the questions. This is an excellent way of reviewing content.

True or false

Level	Beginners and above
Purpose	Active viewing, Reading, Testing
Sequence type	Any
Sequence length	30 seconds to 5 minutes
Activity time	10 minutes

Preparation

Almost any sequence is suitable for this activity. Prepare six statements about the sequence you have chosen. Some should be true and some should be false. You can phrase the statements to pre-teach or reinforce particular language items.

In class

1. Write the statements or make a grid on the board, such as:

	T	*F*
He likes her.		
She likes him.		
He doesn't like her.		
She doesn't like him.		

2. Make sure that the students understand the statements.
3. Tell the students you are going to play a sequence related to the statements. Their task is to watch the sequence and decide whether each statement is true or false.
4. Play the sequence.
5. After viewing, the students have a minute to decide which statements are true and which are false.

6. Elicit and discuss answers from the class.
7. If necessary, play the sequence again to check the answers.
8. The students correct the false statements.
9. Proceed to more intensive study of the sequence, if required.

Variations

■ Beginners and above

Include some statements about which no information is given in the video, and add an 'I don't know' column next to True and False.

■ Intermediate and above

Prepare a narrative summary of the sequence in about ten lines, containing five or more errors of fact (added fragments of information or false statements). Distribute copies of the summary. Students must watch the sequence and decide what's not true. Play the sequence. Afterwards the students read the summary and circle the parts they think are wrong. When they have finished this activity they compare notes with another student. Elicit the false statements in the novelisation from the class. Play the sequence again. The students check their answers.

NOTE: *'True or False' is a very useful testing format.*

Video catch the word

Level Intermediate and above

Purpose Listening

Sequence type Music video

Sequence length 30 seconds (longer if appropriate)

Activity time 30 minutes

Preparation

Select the chorus from a music video (or something similar from a cinema musical or a TV light entertainment show), preferably one which features the actual live performance.

In class

1. Divide the board up into rough word spaces like this:

   — — — — —

   — — — — —

   — — — — —

2. Tell the students that you are going to play part of a music video. Their task to to make out the words.
3. Play the chorus once without stopping.
4. The students must tell you what words they have heard and where they come (which line and the approximate place in the line). Write their words on the board, *even if they are wrong*. Do not expect to elicit all the words at this stage.
5. Play the chorus line by line until the students have given you a complete version.
6. The students read their version of the chorus. Is it correct? Is it consistent? If not, play the chorus again until everyone is satisfied.

7. Read the lyrics aloud. The students compare the version on the board with the actual lyrics. Explain any new words.
8. Discuss the meaning of the lyrics. Play the video again, if necessary, to reinforce the meaning (though a live performance may not help much in this respect).

Variation

The students can start with a verse of the song, although it may be more difficult than the chorus. Alternatively, they can go on to a verse after working out the chorus.

Video dictation

Level Elementary and above

Purpose Listening comprehension

Sequence type Documentary

Sequence length 30–60 seconds

Activity time 10–15 minutes

Preparation

Select a short sequence with clear narration.

In class

1. Tell the students that you are going to play a short documentary sequence three times. Their task is to write down the exact words of the narrator or commentator, but they should not write anything during the first viewing. Explain that you will stop the video several times during the second viewing to give them time to write.
2. Play the sequence through without stopping. Then play it again, pausing after each sentence to allow the students time to write. Finally, play the sequence a third time without stopping.
3. Pairs of students compare what they have written.
4. Student volunteers write sentences from the narration on the board.
5. Ask the class to make any necessary corrections in the sentences on the board.
6. Play the sequence again (e.g. to discuss the subject-matter), and play more of the documentary, if necessary, for other activities.

Voice-over commentary

Level Intermediate and above

Purpose Discussion, Note-taking, Speaking, Written composition

Sequence type Documentary

Sequence length 2–5 minutes

Activity time 30–45 minutes

Preparation

Select a suitable documentary sequence (sports reports and wild life documentaries work especially well). Prepare a transcript of the soundtrack and make enough copies for all the students. Bring an audiocassette recorder and a blank audiocassette to class.

In class

1. Distribute the transcript. Tell the students that you are going to play the corresponding documentary sequence without sound. Their task is to write a new version of the soundtrack which could be followed by students less fluent in English.
2. Play the sequence with the sound *turned down*.
3. Divide the class into three to five groups. Assign one section of the sequence to each group. Tell the groups to plan scripts for their particular sections.
4. Replay the sequence with the sound *turned down*.
5. The groups prepare their scripts.
6. Replay the sequence with the sound *turned down*, pausing or stopping at the end of each section. While watching, groups record their scripts on audiocassette, synchronising sound with the pictures shown on the screen. It may be necessary to rehearse this stage.
7. Replay the sequence and the students' recording together.

8. Replay the sequence with the original soundtrack. Compare this with the students' version.

Variation

■ Advanced: News or sports project

Explain to the students that they are reporters who have received pictures via satellite with no sound. They must write and broadcast a voice-over commentary to go with the pictures. Assign sections of the sequence to groups. Play the sequence with the sound turned down. The groups discuss the subject-matter. Play the sequence again and pause after each change of action or shot. The groups take notes. If possible, they should time each shot. The notes might look like this:

SHOT 1 Duration 10 secs.
Crowd round golf course.
Ballesteros hits the ball.

SHOT 2 Duration 10 secs.
Ball lands in a bunker (sand).

SHOT 3 Duration 5 secs.
Close up of Ballesteros.
No emotion.

Each section of the sequence is played as often as necessary. Each group has 10 minutes to write its commentary. The complete sequence is played once more and each group rehearses its voice-over and then delivers it to picture. Finally the sequence is played with sound, and students can compare their commentaries with the original soundtrack.

> TIP: *If the video recorder has a second audio channel, each group can record its commentary on the video without wiping the master soundtrack. This allows greater authenticity and accuracy at playback.*

> NOTE: *This is one of the most challenging activities in this book. It is also one of the most rewarding. Putting advanced learners under pressure of a deadline and writing to a specific purpose improves their use of language. Doing this task at regular intervals will greatly improve their awareness of word selection, description and writing fluency. It is also helpful in the study of style.*

Watchers and listeners

Level	Intermediate and above
Purpose	Listening, Oral composition, Speaking
Sequence type	Drama
Sequence length	1–3 minutes
Activity time	30 minutes

Preparation

Select a drama sequence or comedy sketch with a clear story-line and a soundtrack that doesn't tell the whole story.

In class

1. Divide the students into pairs. One in each pair is a listener who faces away from the screen. The other is a watcher who faces the screen.
2. Give the task. The watchers must tell the listeners the story after the sequence has been played.
3. Play the sequence.
4. The watchers have 3 minutes to tell the listeners what they saw.
5. Elicit the story from the listeners. They must tell you what they were told. Encourage disagreement: 'Did you hear that?' 'What did you hear?'.
6. At the end, encourage one or two quieter listeners to sum up the story or the disagreements.
7. Replay the sequence. This time all the students watch. At the end the listeners and the watchers compare their earlier versions.

Variation

Instead of the watchers telling the listeners what happened, the listeners ask the watchers questions to find out what happened.

> NOTE: *Each partner in a pair can take up a role appropriate to the subject of the sequence (e.g. detective and witness in a thriller, husband and wife in a situation comedy).*

What did they say?

Level Intermediate and above

Purpose Grammar review

Sequence type Drama

Sequence length 1–3 minutes

Activity time 15–20 minutes

Preparation

Select a sequence with clearly defined conversational exchanges. Write a version of the whole dialogue in reported speech (e.g. 'The frog told the Princess not to cry'). Make copies for all the students.

In class

1. Distribute the reported speech version. Explain that it is a report of the conversation in a sequence that you are going to play. The students' task is to rewrite the report in the form of the actual dialogue they think will be used in the video.
2. Write one or two examples on the board, e.g.:

   The frog told the Princess not to cry. → Frog: Don't cry, Princess.
   The Princess asked the frog what he wanted. → Princess: What do you want?

3. Tell the students to write the dialogue, using direct speech.
4. As the students complete the task, tell them to compare their dialogues with those of another student.
5. Play the sequence, pausing or stopping at the end of each exchange. The students check their versions against the original dialogue.
6. Play the whole sequence without stopping.

What did you see?

Level Beginners and above

Purpose Active viewing, Vocabulary development

Sequence type Any

Sequence length 30 seconds to 5 minutes

Activity time 15 minutes

Preparation

Select a sequence and choose four or five lexical items that relate to or occur in it. Add two or three items that don't, as distractors. Depending on the language level, the items might be concrete things (objects, people), descriptions, processes, actions or abstract things (emotions and concepts). Such considerations will affect the choice of sequence.

In class

1. Write the items on the board or display them on the overhead projector. The students copy them.
2. Make sure that the students understand all the items. Tell them that you are going to play a sequence. Their task is to watch the sequence and circle the items which occur in it or relate to it.
3. Play the sequence.
4. Students work individually, circling the items
5. Elicit from the students the items they have circled. If any disagreement occurs, play the sequence again.
6. As an extension activity, the students make their own sentences using the new words and phrases.

NOTE: *This activity works especially well with advanced learners where technical or specialised vocabulary needs pre-teaching.*

Variation: Elementary and above

■ 'What did you see?' game

The students view the sequence without having seen any items written up. After viewing, groups of three or four make lists of objects, processes, actions, etc. (depending on the subject-matter) which they recall seeing or hearing mentioned. Allow 5 minutes for this. When the time is up, a volunteer from each group writes its list of items on the board. The longest and most accurate list wins.

What gestures did you see?

Level Elementary and above

Purpose Listening, Speaking, Viewing comprehension, Written composition

Sequence type Drama

Sequence length 1–2 minutes

Activity time 10–15 minutes

Preparation

Select a sequence in which characters use a variety of gestures to reinforce the spoken dialogue.

In class

1. Explain to the students that we often use gestures to communicate meaning. Provide an example by holding up a hand, palm away from your face, with the fingers straight up. Ask for volunteers to tell you what the gesture means. Encourage as many different answers as possible. (E.g. 'Wait!', 'Stop!', 'Five!', 'I have the answer!', etc.)
2. Tell the students that you are going to play a sequence without sound, in which a number of different gestures are used. Their task is to observe the gestures and to think of possible meanings for each one.
3. Play the sequence several times, with the sound *turned down*.
4. The students work in groups of three or four, *demonstrating* the different gestures they saw and suggesting possible meanings for each.
5. Volunteers from each group demonstrate one gesture used in the video and suggest possible meanings. Further possible meanings can be elicited from other groups.
6. Play the sequence with the sound *turned up*. The students listen for the actual words accompanying each gesture in the video.

In the next class

7. Working in pairs or small groups, the students write and perform dialogues using at least three of the gestures observed in the video sequence.

What if . . . ?

Level Intermediate and above

Purpose Cross-cultural comparison, Discussion, Speaking

Sequence type Drama

Sequence length 3–5 minutes

Activity time 20–30 minutes

Preparation

Select a sequence which contains a variety of cultural or social features particular to the country or culture involved.

In class

1. Write the following questions on the board:

   If this scene took place in your country, what differences would there be in:
   - *What the characters say?*
   - *What the characters wear?*
   - *What the characters do?*
   - *The story or situation as a whole?*

2. Draw attention to the questions, and tell the students that you are going to play a sequence which takes place in a particular country. Their task is to imagine how the scene would differ if it took place in their own country.

3. Play the sequence.

4. The students work in groups of three or four discussing the questions. If your class is composed of students from different countries, try to have each group represent a variety of cultural backgrounds. Encourage the students to talk about what *would* and *would not* occur during such a scene in their own country.

5. Groups (or individual students, if they are from different countries) report to the whole class.

What's missing?

Level Elementary and above, Children

Purpose Speaking, Viewing comprehension

Sequence type Drama

Sequence length 10–12 minutes

Activity time 20–25 minutes

Preparation

Select a sequence with a strong story-line, which is appropriate for the activity below.

In class

1. Tell the students/children that you are going to show them a story with one important part left out.
2. Play the sequence, omitting one important part, such as how the story starts, or what happens at the end, or a crucial event in the middle.
3. Ask the students/children to guess what has been left out. Elicit answers from them.
4. Play the entire sequence, so that the students/children can see which of their guesses were correct.

What's the conversation?

Level Elementary and above

Purpose Speaking, Viewing comprehension, Written composition

Sequence type Drama

Sequence length 1–2 minutes

Activity time 30 minutes

Preparation

Select a drama sequence in which there are some *visual* clues to the content of the dialogue. It should preferably involve a conversation between two people.

In class

1. Tell the students that you are going to play a sequence without sound, in which a conversation takes place. Their task is to work with a partner and write a dialogue to go with the pictures, and later to act it out in front of the class.
2. Play the sequence one or more times with the sound *turned down.*
3. The students work in pairs (or small groups) and together write a dialogue to go with the pictures.
4. While the pairs or groups are writing their dialogues, circulate among them, answering questions and providing help as needed.
5. As students finish writing, tell them to rehearse the dialogues in their seats.
6. Individual pairs or groups take turns performing their dialogues in front of the class.
7. Play the sequence with sound, to compare the original dialogue with the students' versions.

Variation: Intermediate and above

■ Music video

Choose a music video with a dramatic setting (e.g. a party round a swimming pool). The students imagine what the characters are doing and saying, and in groups write ten-line dialogues representing a scene in that setting. The dialogues can be acted out or read out by the students.

What's the product?

Level Elementary and above

Purpose Discussion, Viewing comprehension

Sequence type TV commercial

Sequence length 30–60 seconds

Activity time 5–10 minutes

Preparation

Select a TV commercial which does not give the identity of the product or service concerned until the end (or near the end).

In class

1. Tell the students that you are going to play part of a TV commercial without sound. Their task is to decide on the product or service being advertised.
2. Play the commercial with the sound *turned down*, stopping just before the identity of the product or service is revealed on the screen.
3. The students work in groups of three or four, working out what is being advertised and giving reasons for their particular choices.
4. Play the complete commercial with the sound *turned up*.
5. Volunteers from groups which guessed correctly tell the class how they were able to know what was being advertised.

Variation

Select at least five commercials and prepare a list of products or services advertised. The students must decide which commercial advertises which product, on the basis of viewing only parts of each commercial.

The activity's effectiveness depends on the commercials' own advertising strategy (i.e. not immediately revealing the nature of the produce or service concerned). With advanced students the activity can thus lead to a discussion of different methods used in advertising on television.

What was the line?

Level Elementary and above

Purpose Listening, Speaking, Vocabulary review, Written composition

Sequence type Drama

Sequence length 1–2 minutes

Activity time 10–15 minutes

Preparation

Select a sequence with appropriate dialogue, so that you can prepare a list of five to ten key words and expressions used in the sequence. Make copies for all the students.

In class

1. Distribute the list.
2. Tell the students that you are going to play a sequence three times, which features each of the key words and expressions in the list. Their task is to listen for each key word or expression, and later recall the line of dialogue in which it was used. Tell the students they are not to take notes during the viewing.
3. Play the sequence three times.
4. The students work in groups of three or four and together try to recall and write down the line of dialogue in which each key word or expression was used.
5. Volunteers go to the board and write the lines of dialogue they have recalled.
6. Play the sequence again. The students check their answers.

Homework

Have students write a new dialogue incorporating three to five lines of the original dialogue from the sequence viewed.

In the next class

The students read their new dialogues to the class.

Where and when?

Level Elementary and above

Purpose Discussion, Viewing comprehension

Sequence type Documentary, Drama

Sequence length 30–60 seconds

Activity time 10–15 minutes

Preparation

Select a sequence which rapidly provides visual or verbal clues to the place and time of the scene or events portrayed, without actually mentioning or showing place names or dates.

In class

1. Explain to the students that you are going to play a very short sequence in which several clues are given to the place and time period of the scene or situation. Their task is to decide where and when the scene takes place (e.g. at a railway station in a small town in the early 1900s).
2. Play the sequence.
3. The students work in groups of three or four and decide together where and when the scene takes place. Tell them to discuss the relevant details they have observed and to give reasons for their answers.
4. One student from each group reports to the class on the details observed and the group's decision as to the place and time of the scene.
5. Play the sequence again.
6. The class vote on which group's account was the most accurate.

Variation: Intermediate and advanced

This activity can form the basis for discussion of social, political or historical issues.

Who's who?

Level	Elementary and above
Purpose	Listening, Note-taking, Speaking, Viewing comprehension
Sequence type	Drama, Music video
Sequence length	3–5 minutes
Activity time	15–20 minutes

Preparation

Select a sequence with at least five characters. Prepare a list of the names of characters appearing in the sequence.

In class

1. Write the names on the board.
2. Tell the students that you are going to play a sequence in which the listed characters appear. Their task is to decide who is who. They may identify a character by providing a physical description (e.g. 'Anna is the tall woman in the red dress') or by stating how the character is related to another character (e.g. 'Helen is Susan's mother'). The students may take brief notes on the characters as they watch.
3. Play the sequence.
4. The students work individually and write a sentence about each character, based on information in the sequence.
5. Play the sequence again, pausing at times if necessary. This time the students check the information in their sentences.
6. The students work in pairs, comparing their sentences.

Variation

■ Describing the characters

Ask students to write a sentence about each character stating one thing (or several things) the character did during the sequence (e.g. 'John answered the telephone').

■ Music video

With a music video get the students to state the number of musicians and to identify what instruments they play. This can lead to a discussion of 'Who's your favourite singer?', 'Who plays bass guitar?', etc.

Words in the news

Level	Intermediate and above
Purpose	Listening, Vocabulary review
Sequence type	TV news programme
Sequence length	4–5 minutes
Activity time	30 minutes

Preparation

Select a typical news summary with predictable items or stories (e.g. the weather, sports, etc.)

In class

1. Tell the students that you are going to play a typical TV summary of local, national or international news. Their task is to predict what kinds of news items are likely to be featured and the vocabulary that will be used in connection with each.
2. Have the students work in pairs, preparing a list of predicted items and vocabulary.
3. Volunteers read examples of items and vocabulary from their lists. As the items are read, write them on the board.
4. Play the news summary.
5. Check the lists on the board against the items and vocabulary which actually occurred in the summary.
6. Play the summary again, pausing or stopping after each news item to answer questions about the item and the vocabulary used.

Word search

Level Intermediate and above

Purpose Note-taking, Viewing comprehension, Vocabulary development, Written composition

Sequence type Drama, Music video, TV news programme

Sequence length 3 minutes

Activity time 15–30 minutes

Preparation

Select a sequence with plenty of striking imagery.

In class

1. Tell the students what kind of sequence you are going to play. Their task is to note down words evoked or suggested by the images they see on the screen.
2. Run the video on fast-forward search so that the students get only an impressionistic view of the sequence.
3. After viewing, the students note down words evoked by the sequence.
4. Divide the students into groups. The members of each group compare and discuss the words they have chosen. Give help with vocabulary if necessary.
5. Each group reconstructs the video story or subject, using the words they have noted down. The reconstructed version should be about ten sentences long and should have a title. (NB: With music videos, this should not be the title of the song but a title which fits the video story or subject-matter.)
6. A spokesperson for each group reads out the group's composition.
7. Play the video at normal speed, more than once if necessary. Which groups got it right? Discuss various versions.

Write a letter

Level Intermediate and above

Purpose Written composition

Sequence type Drama

Sequence length 5–7 minutes

Activity time 30–45 minutes

Preparation

Select a sequence featuring a dramatic or memorable incident.

In class

1. Tell the students the general nature of the sequence you are going to play. Give a brief summary (two or three sentences) of the background information necessary to understand the subject-matter. The students' task is simply to watch what happens.
2. Play the sequence.
3. Tell the students that you are going to play the sequence again. This time, their task is to take notes on what happens so that they can later describe the incident to a friend.
4. Play the sequence again. Pause or stop from time to time and ask questions to guide the students in their note-taking and to elicit appropriate vocabulary.
5. The students imagine they are one of the characters in the sequence. They write the letter which the character would write to a friend, describing the incident portrayed.
6. Play the sequence again for the students to check their letters against the video.

Write a report

Level	Intermediate and above
Purpose	Listening, Note-taking, Viewing comprehension, Written composition
Sequence type	Documentary, Drama, TV news programme
Sequence length	10–20 minutes
Activity time	30–45 minues

Preparation

Select a news, documentary or drama sequence that provides a good basis for an objective report. Decide on a word limit for the report, based on the ability of your students.

In class

1. Tell the students the nature of the sequence you are going to play, (e.g. 'This is a news story about Koko, a talking gorilla'). Their task is to take notes and then write an objective report, in their own words, on the important events and facts presented in the sequence.
2. Play the sequence, pausing or stopping from time to time to allow the students to take notes and ask any questions they may have about content or vocabulary.
3. The students write their reports.
4. Play the sequence again so that the students can check their reports.

Yes, no, maybe

Level Elementary and above

Purpose Discussion, Speaking

Sequence type Drama

Sequence length 2–5 minutes

Activity time 10–20 minutes

Preparation

Select a sequence which is suitable for speculation about motives, feelings and social behaviour. Prepare a list of questions beginning with 'Do you think . . .?' about the characters and the action in the sequence, (e.g. from the film *Casablanca*: 'Do you think Sam was polite toIlsa?', 'Do you think Rick enjoyed seeing Ilsa?', 'Do you think Ilsa is in love with Rick?', etc.). Make copies for all the students.

In class

1. Distribute the list. Tell the students that you are going to play a drama sequence. Their task is to think about their answers to the questions on the list.

2. Play the sequence.

3. Write the following formulae on the board:

   - *I think so, because . . .*
   - *I don't think so, because . . .*
   - *Maybe, because . . .*

4. Give an example by asking two or three of the students the first question and pointing out the formulae they may choose to give their answers.

5. The students work in groups of three or four, discussing the questions and using the formulae in their answers.

6. When the groups have finished, lead a short discussion of the questions, encouraging as many different opinions as possible.
7. Play the sequence again, relating it to the discussion.

APPENDICES

Appendix 1: Logistics

Equipment

Formats and standards

Your equipment must be compatible with the videocassettes you are using. The most common format is VHS, but there are also Betamax and U-matic (a semi-professional ¾-inch format).

Your equipment and videocassettes should also conform to the TV standard adopted in your country. The three main standards are *NTSC* (mainly in North America, parts of Latin America, and Japan and Korea), *SECAM* (in France, USSR and parts of Africa and the Middle East) and *PAL* (in the UK and the rest of the world). See Appendix 2 (p. 156) for a country-by-country breakdown. These differences should not worry you unless you are buying equipment or videocassettes abroad, when you need to make sure you choose the right standard. Moreover, multi-standard equipment which accepts two or three standards is becoming increasingly common. The switches for selecting the relevant standard are clearly marked on the equipment, or the standard is changed automatically to suit the videocassette concerned.

Video recorders

All video recorders or players have certain basic features. However, there are a lot of different makes and models with different characteristics. A particular model may even change significantly from one year to the next. So, get to know the equipment you are going to use or be responsible for.

Essential controls

- *Play and Stop:* These controls operate the video, but on most recorders, when you press the Stop button, the video winds back a few seconds so that you cannot re-start it exactly where you stopped it. Find out how far your own equipment winds back.
- *Fast-forward and Rewind:* These controls merely wind the video forward or back rapidly without showing a picture on the screen.
- *Pause:* The Pause button (sometimes called *Freeze-frame*) freezes the picture on the screen, giving you a still picture frame. On some equipment there is also an *Adjust* button which ensures that there is no interference with the still frame.

Different makes or models have different ways of re-starting the video after a pause (e.g. by pressing Play or by pressing Pause again).

- *Counter:* The counter moves forward and back in time with the video. It can be set at zero at the beginning of a cassette or a sequence to help you find your place. You can also note down counter readings to indicate other significant points in a sequence, and use these readings together with the fast-forward and rewind facility.

  Few counters are really accurate and very few actually indicate minutes or seconds. Those which do are, however, reliable and accurate.

- *Memory:* The memory is linked to the counter. So, if you set the counter at zero at the beginning of a sequence or extract and then press the Memory button, the video will go back to zero (i.e. the beginning of your sequence) every time you rewind.

- *Search:* The search controls allow you to wind forward or back while keeping the picture on the screen. This makes it easier to find your place than using the counter. In fact, you should use a combination of the counter, fast-forward, rewind, search and pause controls to find your place quickly.

  The search facility is achieved by pressing specific controls or by pressing Fast-forward or Rewind while the Play button is depressed.

- *Remote control:* A remote control unit enables you to operate the video recorder without standing next to it. Some units are connected to the equipment by a cord, but most up-to-date ones use an infra-red beam which allows you to stand anywhere within sight of the controls on the recorder.

Optional extras

- *Audio dub:* This socket allows you to connect a recorder or microphone and record another soundtrack to replace the original one. This is very useful if your soundtrack is in L1 and you want a soundtrack in L2. Also, if the language is too advanced for your students, you can re-record it at a simpler level. On some equipment there is a second audio dub socket which allows you to record an additional soundtrack *without* erasing the original one.

- *Sound cut-out:* Some equipment includes a sound cut-out switch which allows you to lose the sound without having to turn down the volume. You regain the sound by pressing the switch again. The sound cut-out facility may be on the remote control unit or on the TV monitor.

TV monitors

TV monitors vary as much as video recorders. Once again, you must get to know the specific features of the equipment you are using. In particular, you need to make sure the monitor is correctly connected and tuned.

- *Connections:* Most video recorders will play into most TV monitors, but many monitors will accept only certain types of connections (partly depending on whether they receive sound and picture through one or more leads/cables). The connections can vary and you may need to consult a technician or the equipment manuals, or both, to get the connections right the first time. Your institution should always have a supply of spare leads/cables with the appropriate connections.
- *Video channel:* The video recorder plays through a specific channel on the TV monitor. This is usually marked *Video* or *AV*, but on some monitors it is necessary to search for the relevant channel.
- *Tuning:* There are several ways in which you can tune equipment to get a better picture or eliminate interference. For example, you can make the picture brighter or adjust the contrast; you can correct the way in which the picture is held horizontally or vertically on the screen; you can get rid of an unsteady picture or interference on the screen by careful use of the *tracking* control located on the video recorder and/or the tuning console. And on most NTSC monitors it is possible to adjust the colour.

You should become familiar with the controls on the console, which is often concealed behind a panel on the monitor. Some of them work within very narrow limits, particularly the tracking control, and only a very slight adjustment may be necessary to correct a fault. Be patient when making such adjustments.

Faults

Here is a simple video fault-finder for your video recorder (VCR) and TV monitor.

Problem	*Check*
No power	VCR and TV connected to mains outlet? Mains switched on? VCR/TV switched on?

No picture	TV switched on? Video lead/cable connected? AV button depressed? TV correctly tuned? Brightness knob turned down? Video channel selected? Video cassette format compatible with equipment?
No sound	Volume control turned down? Sound cut-out button depressed? Sound lead/cable connected?
Interference	Adjust tracking? Adjust tuning? Videocassette standard compatible with equipment?
Unwanted sound	Incorrect tuning? TV 'slipped' off video channel?
No response from switches	Turn off Timer button? Is Operate switch on?
Videocassette won't insert	Remove other videocassette? Turn off Timer button? Videocassette compatible with system?

Where to put the equipment

The ideal — A video recorder under the teacher's control in every classroom.

Unfortunately, this is beyond some institutions' budgets. It also poses security problems.

A video room — A room dedicated to video use, with enough space, enough seats and a big enough screen to accommodate the largest class in the institution. This solution will suit institutions with separate buildings on the same campus or with classrooms on different floors. Classes are usually booked into the video room by the teacher according to a schedule.

The classroom

Various means can be used to house the equipment. It is possible to build a special cupboard on wheels with lockable doors and small holes at the back for the leads/cables; such large cupboards are difficult to vandalise or steal. Alternatively, the video recorder and TV monitor can be mounted on a heavy duty steel trolley above the level of the class to allow easy viewing. The equipment can be chained or bolted to the base of the trolley.

For institutions in which classrooms are on one floor, the video trolley can be wheeled from class to class as needed. It is important that the last teacher to use the equipment should be responsible for its return to the same place of security every day.

More than one video trolley may be necessary in large institutions.

Security

The problem

Video recorders, and to a lesser extent TV monitors, are subject to theft. However, too much security can be self-defeating as it denies access to the equipment to teachers as well as the potential thieves. Here are some suggestions for greater security.

Insurance

It is essential to insure equipment against theft or damage.

Positive identification

It is helpful to have the institution's name or an identification number or symbol engraved into the bodywork of the equipment. A sign saying '*This equipment is positively identifiable*' will help deter thieves.

Locking

If the equipment can be chained or bolted to a trolley, table or wall, then it will be far harder to steal.

Secure rooms

A secure room should be used for housing the equipment at weekends or when the institution is not in use. This could be a video cupboard or a room in the administration area.

Keys

Keys to the video room or to other secure areas should be signed out to named teachers/administrators and kept in a safe place. It's important to make sure that the video room and video keys are always accessible to people responsible, and are not left with an off-duty or on-vacation janitor or caretaker.

Keeping a video library

Most institutions have very little material when they start using video, and gradually build up a collection or library that requires systematic storage and access facilities. Even where it is not possible to have a comprehensive video resource centre, a well-run video library cuts down frustration and encourages more effective use of video in the classroom.

Isolating items

Video material can be used more efficiently if individual units or sequences can be isolated for use on their own. If your institution has access to editing facilities, it is worth copying units or sequences onto separate cassettes, and even grouping these according to level, sequence type, etc.

First make sure that you have the necessary permission or right to make such copies (see Copyright note below).

Storage

Your institution should have at least one cupboard for video storage. It should have plenty of space for worksheets, supporting materials and other documents as well as videocassettes, and it should be near a video recorder for previewing, checking, etc.

- *Identification*
  Always label videocassettes and their boxes or containers. Apart from the title, keep the information on the label to a minimum (e.g. FOLLOW ME Units 1–8, for the BBC teaching video.).

- *Cataloguing*
  It is worth listing materials on the cupboard door and/or on duplicated sheets or computer print-outs which can be updated as required.

  Ideally, there should be two lists: one which lists the main items in the library (e.g. published video courses, authentic videos, etc.); and another which gives details of sequences suitable for use with various activities or recipes. The second list should indicate as precisely as possible where each sequence can be located on the relevant videocassette (e.g. by counter reading, or by time code in the case of some published video courses), and also its level and sequence type.

- *Worksheets*
  For each video there should be a large envelope or a box in the cupboard for worksheets, role cards, teachers notes, etc.
- *Signing out*
  Teachers should always sign out materials so that other teachers will know where they are and who has got them. There could be a list on the video cupboard with the following headings:
  Title Taken by For use in Date out Date due back Returned

Videocassette maintenance and protection

All videocassettes deteriorate with age and use, but you can make them last longer by following these simple suggestions.

Making copies

Provided you have the necessary permission (see Copyright note below), make a safety copy of your video, or keep the original as a safety copy and make other copies for use in class. Remember that no copy will be quite as good as the original, but make sure that you achieve the best quality possible.

Handling

Handle your cassettes with care, and rewind them after use. You may have been told that frequent use of the Pause button will weaken the tape, but there is little evidence to support this claim.

Do not play a cassette if the tape is broken or twisted. It may be possible to remove the tape from its housing in the cassette and untwist it or splice it together. Attempt this only if you have proper splicing material.

Maintenance

Inspect your cassettes frequently for wear and tear, and occasionally play a head-cleaning cassette in the video recorder to clear away any tape deposit.

Avoiding dust

Always keep videocassettes in their cases when not in use, to minimise the risk of carrying dust into the video recorder and damaging it. And the video recorder itself should be kept clean.

Tropicalisation

If you work in a tropical climate, it is worth finding out if you can purchase tropicalised videocassettes which have been specially treated against heat and humidity.

Over-recording

To avoid your video being erased, break off the small plastic square at the back of the cassette. To record over an existing recording, cover the space (where the square was) with tape.

Security

Videocassettes are easier to steal than video recorders, so be sure that your video cupboard is in a secure area.

Copyright

Many institutions and individual teachers are concerned and often bewildered about regulations which might affect their right to use or copy video material for language teaching.

The most important point to remember is that you must abide by the regulations which are legally enforceable in your own country. Beware of advice from outsiders, however well informed they may be. And remember that regulations can change from time to time.

Copying television programmes off air

Regulations vary, but in most countries it is possible to record educational broadcasts for use in class, sometimes subject to the institution concerned holding a licence to copy, and almost always subject to a specified time limit, after which the recordings must be erased. You will probably be allowed to make as many copies as you need.

Beware of the difference between an educational broadcast (i.e. a programme which forms part of an educational broadcast service) and any other broadcast which you may want to use for educational purposes but which may not be covered by provisions relating to educational material. If in doubt, contact the broadcasting organisation concerned.

Pre-recorded educational videos

Videos made and sold for educational purposes (including language learning videos) may not usually be copied without the permission of the distributor and/or producer or publisher concerned. This permission is often withheld, even when it makes sense for an institution to copy selected sequences or make safety copies.

Other pre-recorded videos

In certain countries pre-recorded videos which are not intended for educational use (e.g. feature films, pop videos, etc.) may not be used in class without permission from the distributors or producers concerned. In other countries there are no such restrictions. The same is true of rented videos, most of which are restricted to private viewing in the home.

Copyright notices

The wording of the copyright notices at the beginning of videos varies and is not always clear, and sometimes it has little or no legal validity. Indeed, videos are not always correctly registered for copyright in some countries and may therefore be legally 'in the public domain' without restriction on their use. However, you should not assume that your videos fall into this category without seeking expert legal advice!

Some videos are cleared for educational use only in specified countries. These countries will be indicated in copyright notices on the video or in the accompanying materials.

Audio-visual centres

In some countries videos are available for educational use through specialised audio-visual centres. These centres are subject to specific regulations concerning copying, loaning, etc., which are likely to be different from the regulations which apply to individual institutions or teachers.

Satellite broadcasts

Satellite broadcasts from abroad appear to provide a fruitful source of authentic material in other languages. Unfortunately, their copyright status is still obscure in most countries. You are not likely to be prosecuted if you use a satellite broadcast originated in another country, but you or your institution should watch out for changes in legislation which might define the situation more precisely.

Checklists

Here are two simple checklists to help ensure that your video lessons run smoothly.

Classroom preparation

In preparing for class, have you done the following things:

1. Have you checked the equipment? Is it working?
2. Have you checked your videocassette? It is set up at the right point?
3. Have you checked the sound volume? Go to the back of the room. Is the sound easily audible at the back? Is it reasonably quiet at the front? Remember the sound will be different when the room is full.
4. Have you watched the video from the back of the class? Can everybody see? Do you need a second TV monitor? A larger monitor? Will everybody be able to see enough to take part in the activities which depend on the video?
5. Have you prepared your worksheets and made any last-minute adjustments? You are ready for class if you can answer 'Yes' to the above.

Your first lesson

If you have never used video before and your first lesson is approaching, make sure you have done the following:

1. *Equipment:* Is the equipment set up and running? Test it before going into class.

2. *Instruction:* Has someone demonstrated how to use the equipment (i.e. not just told you but actually demonstrated it to you)?

3. *Help:* Will someone be available (e.g. in the teacher's room or on the phone) in case something goes wrong?

4. *Video:* Have you checked that you've got:
   (a) the videocassette you want?
   (b) a videocassette that works on your equipment?
   (c) the video sequence set up at the place you want to play it?

5. *Worksheets:* Have you prepared your worksheets or teaching exercises and made sufficient copies for the class?

6. *Class layout:* Have you checked that:
   (a) everyone can see the screen well enough to do comprehension tasks?
   (b) everyone can hear the soundtrack?

7. *At the end of the lesson:* Have you:
   (a) retrieved your video from the recorder?
   (b) switched everything off?
   (c) picked up spare materials and left the place clear for the next user?

Appendix 2: TV standards around the world

Country	Lines/ Field	Colour	Voltage (V)	Frequency (Hz)
Abu Dhabi	625/50	PAL	240	50
Afganistan	625/50	PAL	220	50
Alaska	525/60	NTSC	120/240	60
Albania	625/50	SECAM(H)	220	50
Algeria	625/50	PAL	127/220	50
Andorra	625/50	(PAL)	220	50
Angola	625/50		220	50
Antigua & Barbuda	525/60	NTSC	230	60
Antilles	525/50	NTSC	127/220	50/60
Argentina	625/50	PAL	220/225	50
Australia	625/50	PAL	240	50
Austria	625/50	PAL	220	50
Azores	625/50	PAL	110-220	50
Bahamas	525/60	NTSC	120/240	60
Bahrain	625/50	PAL	230/110	50/60
Bangladesh	625/60	PAL	220/230	50
Barbados	525/60	NTSC	115/220	50
Belgium	625/50	PAL	127/220	50
Benin			220	50
Bermuda	525/60	NTSC	120/240	60
Bolivia	525/60	NTSC	115/230	50
Botswana	625/50	PAL	220	50
Brazil	525/60	PALM	220	50/60
Brunei	625/50	PAL	230	50
Bulgaria	625/50	SECAM(V)	220	50
Burma	525/60	NTSC	230	50
Burundi			220	50
Cambodia			120/220	50
Cameroon			127/220	50
Canada	525/60	NTSC	120/240	60
Canary Islands	625/50	PAL	127/220	50
Cayman Islands	525/60	(NTSC)	120/240	60
Central African Empire			220	50
Chad			220	50
Chile	525/60	NTSC	220	50
China	625/50	PAL	220	50
Columbia	525/60	NTSC	120/240	60
Congo	625/50	SECAM(V)	220	50
Costa Rica	525/60	NTSC	120	60
Cuba	525/60	NTSC	115/120	60
Curacao & Aruba	525/60	NTSC	127/220	50
Cyprus	625/50	PAL/SECAM	230	50
Czechoslovakia	625/50	SECAM(V)	220	50
Dahomey			220	50
Denmark	625/50	PAL	220	50

Country	Lines/ Field	Colour	Voltage (V)	Frequency (Hz)
Diego Garcia	525/60	NTSC		
Djibouti	625/50	SECAM	220	50
Dominica			230	50
Dominion Republic	525/60	NTSC	110	60
Dubai	625/50	PAL	220	50
Ecuador	525/60	NTSC	120/127	60
Egypt	625/50	SECAM	220	50
El Salvador	525/60	NTSC	115/230	60
Ethiopia			220	50
Falkland Islands			230	50
Fernando Po			220	50
Fiji			240	50
Finland	625/50	PAL	220	50
France	625/50	SECAM(V)	127/220	50
Gabon	625/50	SECAM(V)	220	50
Gambia			230	50
Germany (East)	625/50	SECAM(V)	127/220	50
Germany (West)	625/50	PAL	220	50
Ghana	625/50	PAL	230/250	50
Gibraltar	625/50	PAL	240	50
Greece	625/50	SECAM(H)	220	50
Greenland	525/60	NTSC	220	50
Grenada			230	50
Guadeloupe	625/50	SECAM(H)	220	50
Guatemala	525/60	NTSC	120/240	60
Guinea-Bissau				
Guinea Equatorial				50
Guinea Republic				50
Guyana French	625/50	SECAM(V)	127/220	50
Guyana Republic			110/220	50
Haiti	625/50	SECAM(V)	220	60
Hawaii	525/60	NTSC	120/240	60
Honduras			110	60
Hong Kong	625/50	PAL	200	50
Hungary	625/50	SECAM(V)	220	50
Iceland	625/50	PAL	220	50
India	625/50	PAL	230/250	50
Indonesia	625/50	PAL	127/220	50
Iran	625/50	SECAM/PAL	220	50
Iraq	625/50	SECAM(H)	220	50
Ireland	625/50	PAL	220	50
Israel	625/50	PAL	230	50
Italy	625/50	PAL	127/220	50
Ivory Coast	625/50	SECAM(V)	220	50
Jamaica			110/220	50
Japan	525/60	NTSC	100/200	60
Jibuti	625/50	SECAM	220	50
Johnston Island	525/60	NTSC		
Jordan	625/50	PAL	220	50

Country	Lines/ Field	Colour	Voltage (V)	Frequency (Hz)
Kampuchea			120/208	50
Kenya	625/50	PAL	240	50
Korea (North)	625/50	SECAM		
	525/60	NTSC	220	60
Korea (South)	525/60	NTSC	100	60
Kuwait	625/50	PAL	240	50
Laos			220	50
Lebanon	625/50	SECAM(V)	110/220	50
Leeward Islands			230	60
Lesotho			220	50
Liberia	625/50	PAL	120/240	60
Libya	625/50	PAL	127/230	50
Liechtenstein			220	50
Luxembourg	625/50	PAL/SECAM	127/220	50
Macau			110/220	
Madagascar	625/50	PAL	127/220	50
Madeira	625/50	PAL	220	50
Malawi			230	50
Malaysia	625/50	PAL	230/240	50
Maldives	625/50	PAL		50
Mali			220	50
Malta	625/50	PAL	240	50
Martinique	625/50	SECAM	127/220	50
Mauritania			220	50
Mauritius	625/50	SECAM(V)	230	50
Mexico	625/50	NTSC	127/220	60
Micronesia	525/60	NTSC		
Midway Island	525/66	NTSC		
Monaco	525/60	SECAM/PAL	127/220	50
Mongolia	625/50	SECAM(V)		50
Montserrat			230	60
Morocco	625/50	SECAM(V)	110/127	50
Mozambique	625/50	PAL	220	50
Namibia	625/50	PAL	220	50
Nauru			240	50
Nepal			230	50
Netherlands	625/50	PAL	220	50
New Caledonia	625/50	SECAM(V)	220	50
New Guinea			240	50
New Zealand	625/50	PAL	230	50
Nicaragua	525/60	NTSC	120/240	60
Niger			220	50
Nigeria	625/50	PAL	220/230	50
Norway	625/50	PAL	230	50
Oman	625/50	PAL	240	50
Pakistan	625/50	PAL	230	50
Panama	525/60	NTSC	120/220	60
Papua New Guinea			240	50
Paraguay	625/50	PALM	220	50
Peru	525/60	NTSC	220/225	60
Philippines	525/60	NTSC	120/220	60

Country	Lines/ Field	Colour	Voltage (V)	Frequency (Hz)
Poland	625/50	SECAM(V)	220	50
Portugal	625/50	PAL	220	50
Puerto Rico	525/60	NTSC	120/240	60
Qatar	625/50	PAL	240	50
Reunion	625/50	SECAM(V)	220	50
Romania	625/50	SECAM	220	50
Rwanda			220	50
Sabah & Sarawak	625/50	PAL	240	50
Samoa Eastern	525/60	NTSC	230	50
Samoa Western			230	50
San Marino	(625/50)	(PAL)	127/220	
Saudi Arabia	625/50	SECAM(H)	127/220	50
Senegal	625/50	SECAM(V)	127	50
Seychelles			230	50
Sierra Leone	625/50	PAL	230	50
Singapore	625/50	PAL	230	50
Solomon Islands			240	50
Somalia			110/230	50
South Africa	625/50	PAL	220/250	50
Spain	625/50	PAL	127/220	50
Sri Lanka	625/50	PAL	230	50
St Helena			230	
St Kitts	525/60	NTSC	230	60
St Lucia			240	50
St Pierre et Miquelon	625/50	SECAM(V)	115	50
St Vincent			230	50
Sudan	525/60	PAL	240	50
Surinam	625/50	NTSC	115/127	60
Swaziland	625/50	PAL	230	50
Sweden	625/50	PAL	220	50
Switzerland	625/50	PAL	220	50
Syria	625/50	SECAM(H)	115/220	50
Tahiti	625/50	SECAM(H)	220	60
Taiwan	525/60	NTSC	110/220	60
Tanzania	625/50	PAL	230	50
Thailand	625/50	PAL	220	50
Tibet			220	50
Togo	625/50	SECAM	127/220	50
Tongo			240	50
Trinidad and Tobago	525/60	NTSC	115/230	60
Trust Islands	525/60	NTSC	110	60
Tunisia	625/50	SECAM(V)	127/220	50
Turkey	625/50	PAL	220	50
Uganda	625/50	PAL	240	50
United Arab Emirates	625/50	PAL	220/240	50
United Kingdom	625/50	PAL	240/220	50
Upper Volta			220	50
Uruguay	625/50	PALM	220	50
United States	525/60	NTSC	120/208	60

Country	Lines/ Field	Colour	Voltage (V)	Frequency (Hz)
USSR	625/50	SECAM(V)	127/220	50
Vatican	(625/50)	(PAL)	127/220	50
Venezuela	525/60	NTSC	120/240	60
Vietnam			127/220	50
Virgin Islands	525/60	NTSC	110/220	60
Yeman Arab Republic	625/50	PAL	220	50
Yeman Democratic Republic	625/50	PAL	230/250	50
Yugoslavia	625/50	PAL	220	50
Zaire	625/50	SECAM(V)	220	50
Zambia	625/50	PAL	220	50

Appendix 3: Recommended reading

Allan, Margaret, *Teaching English with Video*, Longman: London, 1986.
Geddes, Marion, and Sturtridge, Gill, *Use of Video in Language Learning*, Heinemann: London, 1982.
Harris, M., McKillop, J., Maratos, C. and Palmer, N., *Improve Your English Through Television*, Adult Literacy and Basic Skills Unit: London, 1986.
Lavery, Mike, *Active Viewing Plus*, Modern English Publications: London, 1984.
Lonergan, Jack, *Video in Language Teaching*, Cambridge University Press: Cambridge, 1984.
Rinvolucri, Mario, and Lavery, Mike, *Resource Books for Teachers: Video*, Oxford University Press: Oxford, forthcoming.
Tomalin, Barry, *Video, TV and Radio in the English Class*, Macmillan: London, 1986.
Tomalin, Barry, *Video in the English Class* (a training video), BBC English: London, 1990.
Willis, Jane, *Television English*, BBC Publications: London, 1987.

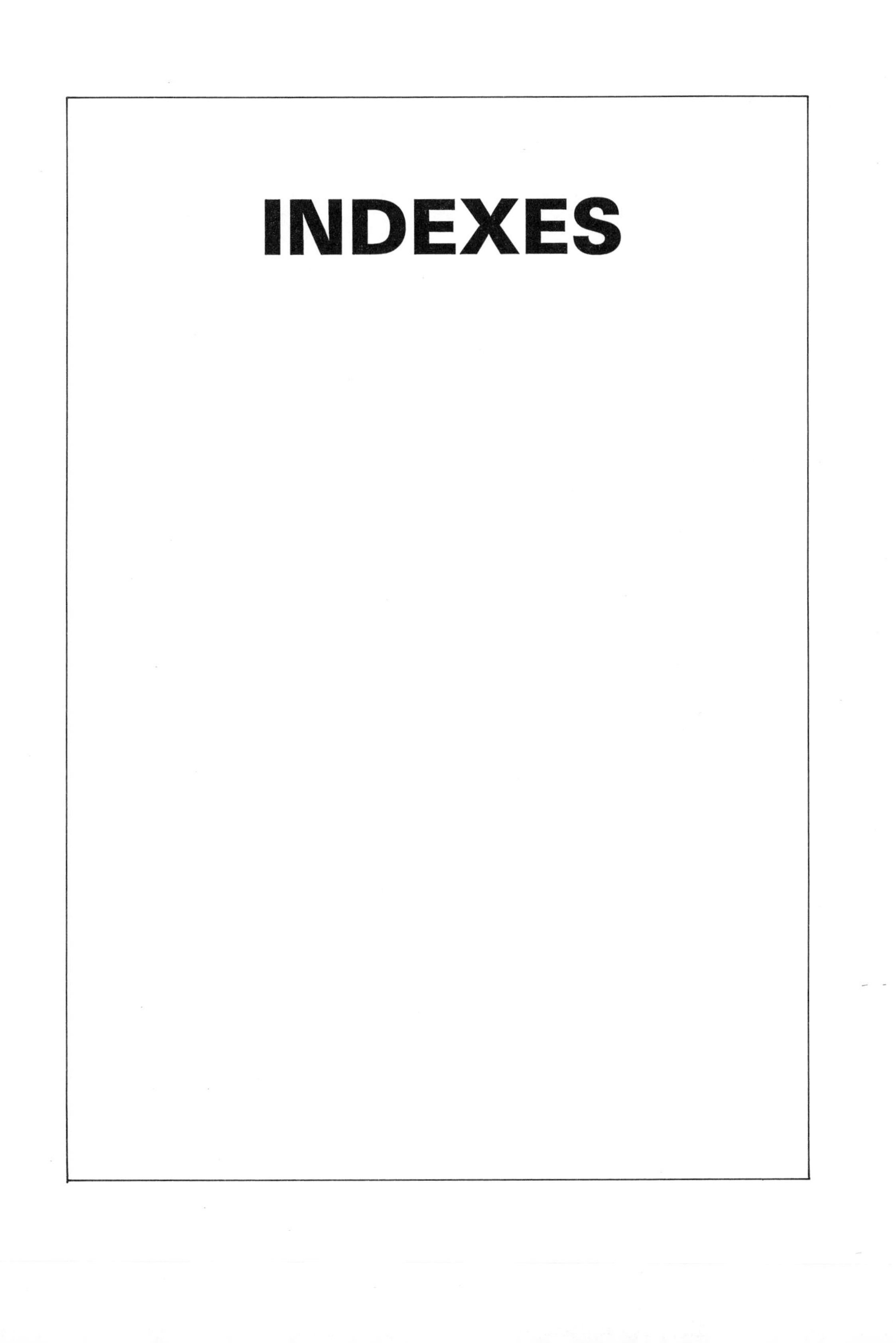

INDEXES

Index of recipes

Index of activities: Level, purpose and sequence type

(NB Activities are listed under the lowest level at which they may be used.)

Level

Children

Beginners

Elementary

Intermediate

Advanced

Purpose

Active viewing

Cross-cultural comparison

Cultural awareness

Discussion

Sequence type